Undergraduate Texts in Mathematics

Editors

S. Axler
F.W. Gehring
K.A. Ribet

Springer

New York
Berlin
Heidelberg
Hong Kong
London
Milan
Paris
Tokyo

Undergraduate Texts in Mathematics

Abbott: Understanding Analysis.

Anglin: Mathematics: A Concise History and Philosophy.
Readings in Mathematics.

Anglin/Lambek: The Heritage of Thales.
Readings in Mathematics.

Apostol: Introduction to Analytic Number Theory. Second edition.

Armstrong: Basic Topology.

Armstrong: Groups and Symmetry.

Axler: Linear Algebra Done Right. Second edition.

Beardon: Limits: A New Approach to Real Analysis.

Bak/Newman: Complex Analysis. Second edition.

Banchoff/Wermer: Linear Algebra Through Geometry. Second edition.

Berberian: A First Course in Real Analysis.

Bix: Conics and Cubics: A Concrete Introduction to Algebraic Curves.

Brémaud: An Introduction to Probabilistic Modeling.

Bressoud: Factorization and Primality Testing.

Bressoud: Second Year Calculus.
Readings in Mathematics.

Brickman: Mathematical Introduction to Linear Programming and Game Theory.

Browder: Mathematical Analysis: An Introduction.

Buchmann: Introduction to Cryptography.

Buskes/van Rooij: Topological Spaces: From Distance to Neighborhood.

Callahan: The Geometry of Spacetime: An Introduction to Special and General Relativity.

Carter/van Brunt: The Lebesgue–Stieltjes Integral: A Practical Introduction.

Cederberg: A Course in Modern Geometries. Second edition.

Childs: A Concrete Introduction to Higher Algebra. Second edition.

Chung/AitSahlia: Elementary Probability Theory: With Stochastic Processes and an Introduction to Mathematical Finance. Fourth edition.

Cox/Little/O'Shea: Ideals, Varieties, and Algorithms. Second edition.

Croom: Basic Concepts of Algebraic Topology.

Curtis: Linear Algebra: An Introductory Approach. Fourth edition.

Daepp/Gorkin: Reading, Writing, and Proving: A Closer Look at Mathematics.

Devlin: The Joy of Sets: Fundamentals of Contemporary Set Theory. Second edition.

Dixmier: General Topology.

Driver: Why Math?

Ebbinghaus/Flum/Thomas: Mathematical Logic. Second edition.

Edgar: Measure, Topology, and Fractal Geometry.

Elaydi: An Introduction to Difference Equations. Second edition.

Erdős/Surányi: Topics in the Theory of Numbers.

Estep: Practical Analysis in One Variable.

Exner: An Accompaniment to Higher Mathematics.

Exner: Inside Calculus.

Fine/Rosenberger: The Fundamental Theory of Algebra.

Fischer: Intermediate Real Analysis.

Flanigan/Kazdan: Calculus Two: Linear and Nonlinear Functions. Second edition.

Fleming: Functions of Several Variables. Second edition.

Foulds: Combinatorial Optimization for Undergraduates.

Foulds: Optimization Techniques: An Introduction.

(continued after index)

Ulrich Daepp   Pamela Gorkin

Reading, Writing, and Proving

A Closer Look at Mathematics

 Springer

Ulrich Daepp
Department of Mathematics
Bucknell University
Lewisburg, PA 17837
USA
udaepp@bucknell.edu

Pamela Gorkin
Department of Mathematics
Bucknell University
Lewisburg, PA 17837
USA
pgorkin@bucknell.edu

Mathematics Subject Classification (2000): 00-01, 03-01

Library of Congress Cataloging-in-Publication Data
Daepp, Ulrich.
    Reading, writing, and proving : a closer look at mathematics / Ulrich Daepp, Pamela
Gorkin.
        p. cm.– (Undergraduate texts in mathematics)
    Includes bibliographical references and index.
    ISBN 0-387-00834-9 (alk. paper)
        1. Mathematics–Study and teaching (Higher)–United States.   2. Technical
    writing–Study and teaching (Higher)–United States.   I. Gorkin, Pamela.   II. Title.
    III. Series.
    QA113.D34   2003
    510-dc21   2003045420

ISBN 0-387-00834-9   Printed on acid-free paper.

Printed in the United States of America.

9 8 7 6 5 4 3 2 1   SPIN 10920411

www.springer-ny.com

Springer-Verlag   New York Berlin Heidelberg
A member of BertelsmannSpringer Science + Business Media GmbH

For Hannes and Madeleine

Preface

You are probably about to teach or take a "first course in proof techniques," or maybe you just want to learn more about mathematics. No matter what the reason, a student who wishes to learn the material in this book likes mathematics, and we hope to keep it that way. At this point, students have an intuitive sense of why things are true, but not the exposure to the detailed and critical thinking necessary to survive in the mathematical world. We have written this book to bridge this gap.

In our experience, students beginning this course have little training in rigorous mathematical reasoning; they need guidance. At the end, they are where they should be; on their own. Our aim is to teach the students to read, write, and do mathematics independently, and to do it with clarity, precision, and care. If we can maintain the enthusiasm they have for the subject, or even create some along the way, our book has done what it was intended to do.

Reading. This book was written for a course we teach to first and second year college students. The style is informal. A few problems require calculus, but these are identified as such. Students will also need to participate while reading proofs, prodded by questions (such as, "Why?"). Many detailed examples are provided in each chapter.

Since we encourage the students to draw pictures, we include many illustrations as well. Exercises, designed to teach certain concepts, are also included. These can be used as a basis for class discussion, or preparation for the class. Students are expected to solve the exercises before moving on to the problems. Complete solutions to almost all of the exercises are provided at the end of each chapter. Problems of varying degrees of difficulty appear at the end of each chapter. Some problems are simply proofs of theorems that students are asked to read and summarize; others supply details to statements in the text. Though many of the remaining problems are standard, we hope that students will solve some of the unique problems presented in each chapter.

Writing. The bad news is that it is not easy to write a proof well. The good news is that with proper instruction, students quickly learn the basics of writing. We try to write in a way that we hope is worthy of imitation, but we also provide students with "tips" on writing, ranging from the (what should be) obvious to the insider's preference ("Don't start a sentence with a symbol.").

Proving. How can someone learn to prove mathematical results? There are many theories on this. We believe that learning mathematics is the same as learning to play an instrument or learning to succeed at a particular sport. Someone must provide the background: the tips, information on the basic skills, and the insider's "know how." Then the student has to practice. Musicians and athletes practice hours a day, and it's not surprising that most mathematicians do, too. We will provide students with the background; the exercises and problems are there for practice. The instructor observes, guides, teaches and, if need be, corrects. As with anything else, the more a student practices, the better she or he will become at solving problems.

Using this book. What should be in a book like this one? Even a quick glance at other texts on this subject will tell you that everyone agrees on certain topics: logic, quantifiers, basic set theoretic concepts, mathematical induction, and the definition and properties of functions. The depth of coverage is open to debate, of course. We try to cover logic and quantifiers fairly quickly, because we believe that students can only fully appreciate the fundamentals of mathematics when they are applied to interesting problems.

What is also apparent is that after these essential concepts, everyone disagrees on what should be included. Even we prefer to vary our approach depending on our students. We have tried to provide enough material for a flexible approach.

- *The Minimal Approach.* If you need only the basics, cover Chapters 1–17. (If you assume the well ordering principle, or decide to accept the principle of mathematical induction without proof, you can also omit Chapter 12.)
- *The Usual Approach.* This approach includes Chapters 1–17 and Chapters 20–22. (This is easily doable in a standard semester, if the class meets three hours per week.)
- *The Algebra Approach.* For an algebraic slant to the course, cover Chapters 1–17 and Chapters 25 and 26.
- *The Analysis Approach.* For a slant towards analysis, cover Chapters 1–22. (This is what we usually cover in our course.) Include as much material from Chapters 23 and 24 as time allows. Students usually enjoy an introduction to metric spaces.
- *Projects.* We have included projects intended to let students demonstrate what they can do when they are on their own. We indicate prerequisites for each project, and have tried to vary them enough that they can be assigned throughout the semester. The results in these projects come from different areas that we find particularly interesting. Students can be guided to a project at their level. Since there are open-ended parts in each project, students can take these projects as far as they want to. We usually encourage the students to work on these in groups.
- *Notation.* A word about some of our symbols is in order here. In an attempt to make this book user-friendly, we indicate the end of a proof with the well-known symbol ■. The end of an example or exercise is designated by ○. If a problem is used later in the text, we designate it by **Problem**♮. We also have a fair number of "non-proofs." These are proofs that are questionable, and students are asked to find the error. We conclude such proofs with the symbol ⍰. Every other symbol will be defined when we introduce you to

it. Definitions are incorporated in the text for ease of reading and the terms defined are given in bold-face type.

Presenting. We also hope that students will make the transition to thinking of themselves as members of a mathematical community. We encourage the students we have in this class to attend talks, give talks, go to conferences, read mathematical books, watch mathematical movies, read journal articles, and talk with their colleagues about the things in this course that interest them. Our (incomplete, but lengthy) list of references should serve a student well as a starting point. Each of the projects works well as the basis of a talk for students, and we have included some background material in each section. We begin the chapter on projects with some tips on speaking about mathematics.

We hope that through reading, writing, proving, and presenting mathematics, we can produce students who will make good colleagues in every sense of the word.

* * *

Acknowledgments. Writing a book is a long process, and we wish to express our gratitude to those who have helped us along the way. We are, of course, grateful to the students at Bucknell University who suffered through the early versions of the manuscript, as well as those who used later versions. Their comments, suggestions, and detection of errors are most appreciated. We thank Andrew Shaffer for help with the illustrations. We also wish to express our thanks to our colleagues and friends, Gregory Adams, Thomas Cassidy, David Farmer, and Paul McGuire for helpful conversations. We are particularly grateful to Raymond Mortini for his willingness to carefully read (and criticize) the entire text. The book is surely better for it. We also wish to thank our (former) student editor, Brad Parker. We simply cannot overstate the value of Brad's careful reading, insightful comments, and his suggestions for better prose. We thank Universität Bern, Switzerland, for support provided during our sabbaticals. Finally, we thank Hannes and Madeleine Daepp for putting up with infinitely many dinner conversations about this text.

* * *

We plan to maintain a website with additional material, corrections, and other documentation at

http://www.facstaff.bucknell.edu/udaepp/readwriteprove/

Ulrich Daepp and Pamela Gorkin
Lewisburg, Pennsylvania 2003

udaepp@bucknell.edu
pgorkin@bucknell.edu

Contents

CHAPTER 1

The How, When, and Why of Mathematics

What is mathematics? Many people think of mathematics (incorrectly) as addition, subtraction, multiplication, and division of numbers. Those with more mathematical training may think of it as dealing with algorithms. But most professional mathematicians think of it as much more than that. While we certainly hope that our students will perform algorithms correctly, what we really want is for them to understand three things: how you do something, why it works, and when it works. The problems we present to you in this book concentrate on these three goals. If this is the first time you have been asked to prove theorems, you may find this to be quite a challenge. Not only will you be learning how to solve the problem, you will also be learning how to write up the solution. The necessary definitions and background to understand a problem, as well as a general plan of attack, will always be presented in the text. It's up to you to spend the time reading, trying various approaches, rereading, and reapproaching. You will probably be spending more time on fewer exercises than you ever have before. While you are now beyond the stage of being given steps to follow and practice, there are general rules that can assist you in your transition to doing higher mathematics. Many people have written about this subject before.

The classic text on how to approach a problem is a wonderful book called *How to Solve It* by George Pólya, [66].

In his text, Pólya gives a list of guidelines for solving mathematical problems. He calls his suggestions "the list." We have included the original in Appendix 28.3. This list has served as a guide for several generations of mathematicians, and we suggest that you let it guide you as well. Here's a closer look at "the list" with some 21st-century modifications.

First. "Understanding the problem." Easier said than done, of course. What should you do? Make sure you know what all the words mean. You may need to look something up in this book, or you may need to use another book. Look at the statement to figure out carefully what you are given and what you are supposed to figure out. If a picture will help, draw it. Will you be proving something? What? Will you have to obtain an example? Of what? Check all conditions. Will you have to show that something is false? Once you understand what you have to do, you can move on to the next step.

Second. "Devising a plan." How will you attack the problem? At this point, you understand what must be done (because you have completed Step 1). Have you seen something like it before? If you haven't looked over class notes, haven't read the text, or haven't done the previous homework assignments, the odds are slim that you have seen anything that will be helpful. Do all that first. Look over the text with the problem in mind, read over your notes with the problem in your head, look at previous exercises and theorems that sound similar. Maybe you can use some of the ideas in the proof of a theorem, or maybe you can use a previous homework problem. Mathematics builds on itself and the problems in the text will also. If you are truly stuck, try to answer a simpler, similar question. Once you decide on a method of approach, try it out.

Third. "Carrying out the plan." Solve the problem. Look at your solution. Is each sentence true? Sometimes it is difficult to catch an error right after you have "found a solution." Put the problem down and come back to it a few hours later. Is each sentence still true?

Fourth. "Looking Back." Pólya suggests checking the result and the argument, or even looking for a different proof. If you are allowed (check with your teacher), one really good way to check a proof is to give it to someone else. You can present it to friends. Even if they don't understand a word you are saying, sometimes saying it out loud in a coherent manner will allow you to recognize an error you can't spot when you are reading. If you are permitted to work together, switch proofs and ask your partner for criticism of your proof.

When you are convinced that your argument is correct, it is time to write up a correct and neat solution to the problem.

Here is an example of the Pólya method at work in mathematics; we will decipher a message. A cipher is a system that is used to hide the meaning of a message by replacing the letters of the alphabet by other letters or symbols.

Exercise 1.1.

The following message is encoded by a shift of the alphabet; that is, every letter is replaced by another one that has been shifted n places further down the alphabet. Once we reach the end of the alphabet, we start over. For instance, if n were 7, we would make the replacements a $\rightarrow$ h, b $\rightarrow$ i, ..., s $\rightarrow$ z, t $\rightarrow$ a, ... Now the exercise: What does the message below say?

PDEO AJYKZEJC WHCKNEPDI EO YWHHAZ W YWAOWN YELDAN. EP EO RANU AWOU PK XNAWG, NECDP?

Let's use the ideas from Pólya's list to solve this. If you have solved problems like this before, it might be a better exercise for you to try on your own to see how this fits Pólya's method before you read on.

1. *"Understanding the problem."* Each sequence of letters with no blank space between the letters represents one word. Each letter is shifted by the same number of places: namely n. So n is the unknown in this problem and it is what we need to find. Once we know the value of n, we can decipher the whole message. In addition, once we know the meaning of one letter, we can find the value for n.

2. *"Devising a plan."* A cipher text may have weak points. What are these? How about the short words? Looking at the short words, in some sense, substitutes an easier problem for the one we have.

3. *"Carrying out the plan."* The short words are:

   W;

   EO (which appears twice);

   EP;

   PK.

   Try using the most common one and two letter words. For each guess, check the beginning of the cipher text to see if it makes sense. It shouldn't take long for you to come up with the message.
4. *"Looking Back."* If your solution makes sense, then it is highly unlikely that a different replacement is also possible. So the solution is (with high probability) the only one.

   Would there have been other solution methods? Sure. For instance, not all letters have the same frequency in the English language. One analysis of English texts showed the letter e occurring most frequently, followed by (in this order) t, a, o, i, n, s, h, and r. (See [78, p. 19].) We could have used this information to guess the assignment of letters.

   We also could have simply tried one value of n after another until the message made sense.

Have you now solved the problem? If you know what the message says, then the answer to this question is *yes*. Are you done? Unless you solved the problem and wrote up a clear, complete solution, the answer to this second question is *no*. A solution consists of a report that tells the reader how you solved the problem and what the answer is. This needs to be done in clear English sentences. As you write up your solution, try to keep the reader in mind. You should explain things clearly and logically, so that the reader doesn't have to spend time filling in gaps.   ○

We now move on to a very different kind of example. Consider the set of points in three-space. In case you haven't seen this before, these points are easily described. We take the familiar xy-plane, and place it parallel to the floor. The z-axis is the vertical line perpendicular to the xy-plane and passing through the origin of the xy-plane (see Figure 1.1).

We'll review the important concepts before we begin our example.

FIGURE 1.1

To locate a point, we will give three coordinates. The first coordinate is the x-coordinate and tells us the number of units to walk in the x-direction. The second is the y-coordinate, telling us how to move in the y-direction and the third is the z-coordinate, telling us how far, up or down, to move. So a point in three-space is denoted by (x, y, z). It is important to make sure you understand this. Try to think of how you would plot points. The point $(1, 0, 0)$ (plotted in Figure 1.2) would appear one unit in the positive direction on the x-axis (since it doesn't move in the y-direction or z-direction at all). The point $(-1, 1, 0)$ would appear in the xy-plane, one unit back on the x-axis and one unit in the positive y-direction. Finally the point $(2, -1, 3)$ is plotted in Figure 1.2.

Let's go a bit further here. In two-space, what was $x = 0$? Since y does not appear in that equation, it is unrestricted and can be any real number. That's why $x = 0$ in two-space is the y-axis. What is $x = 3$? It is a line parallel to the y-axis through the point $(3, 0)$. So, let's try to generalize this to the situation in three-space. What's the plane $z = 0$? Recall that if a variable doesn't appear, then it may assume any value. So this means that z is fixed at 0 while x can take any value, as can y. Thus, the plane $z = 0$ is the xy-plane. Similarly, the yz-plane is the plane $x = 0$ and the xz-plane is the plane $y = 0$. These three planes are called the coordinate planes. What's the plane $z = 3$? $x = 2$? $y = y_0$? There's plenty to think about here, but let's start by asking what the distance is between two points in three-space.

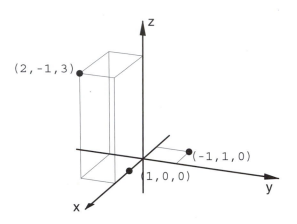

FIGURE 1.2

Example 1.2.
Given two points (x_0, y_0, z_0) and (x_1, y_1, z_1) in three-space, what is the distance between the two points?

We follow Pólya's method to find the solution.

1. *"Understanding the problem."* Before we begin, we make sure we really understand the meaning of each word and symbol above. We spent the last few paragraphs making sure we all understand the symbols, and all the words are familiar ones that appear in a standard English dictionary. But, wait—has "distance between two points" really been defined? We need to be sure that everyone means the same thing by this. The distance between these two arbitrary points would mean the length of the straight line segment joining the two points. That's what we need to find. What were we given? Two points and their coordinates.

2. *"Devising a plan."* How do we solve something like this? We haven't covered anything yet, so what can the authors be thinking? If you have no idea how to get started, try thinking about finding the distance between two specific points. Of course, (and this is very important) this won't give us a general formula because it is much too specific, but maybe we'll get some ideas.
So what's the distance between the two points $(1, 0, 0)$ and $(-1, 0, 0)$? That question is easier to answer—it's two. What's

the distance between $(1, 1, 0)$ and $(-1, -2, 0)$? This seems to be just the distance between two points in the familiar xy-plane. We saw a formula for that at some point. It was obtained using the Pythagorean Theorem. What was it? If you can't recall the formula, look it up or (better, yet) try to derive it again.

Our reasoning now brings us to a simpler, similar question. As you recall, this is precisely where Pólya suggested we look for a plan. So far, it seems we can find the distance between two points as long as they lie in a plane parallel to one of the coordinate planes. But in this problem, if we look at the two points, they need not lie in such a plane. We can try to insert a third point that helps us to reduce the problem to one we can already solve. Which point? A picture will help here, so we draw one in Figure 1.3.

We see that (x_0, y_0, z_0) and (x_1, y_1, z_0) lie in the plane $z = z_0$, while (x_1, y_1, z_0) and (x_1, y_1, z_1) lie on the same vertical line, in the intersection of the two planes, $x = x_1$ and $y = y_1$. We "devise our plan" using these three points. Can we get the distance we are looking for from these three points? Look at Figure 1.3 and see if you can guess the rest before going on to Step 3. You probably noticed that the vertical line makes a right angle with every line in the plane $z = z_0$. This should suggest something to you—something like the Pythagorean Theorem.

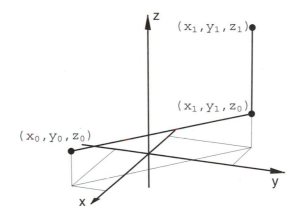

FIGURE 1.3

3. *"Carrying out the plan."* This is the only thing the reader will see. Everything that preceded this was to assist us in obtaining this solution. That means the reader doesn't know what the points are; we have to tell him or her that. We should make sure we say why a sentence follows from the previous one and we should use equal signs between equal objects. When we think we are done, we should tell the reader that too.

Solution.
Let $P = (x_0, y_0, z_0)$ and $Q = (x_1, y_1, z_1)$ be two points in space. We claim that the distance between these two points, denoted by $d(P, Q)$, is

$$d(P, Q) = \sqrt{(x_0 - x_1)^2 + (y_0 - y_1)^2 + (z_0 - z_1)^2}.$$

Proof.
We introduce a third point with coordinates $R = (x_1, y_1, z_0)$. Since (x_0, y_0, z_0) and (x_1, y_1, z_0) both lie in the plane $z = z_0$, we can use the distance formula for two points in a plane to find the distance between them. Thus, the distance is given by

$$d(P, R) = \sqrt{(x_0 - x_1)^2 + (y_0 - y_1)^2}.$$

Now look at the distance between the two points (x_1, y_1, z_0) and (x_1, y_1, z_1). Since these points lie on the same vertical line, the distance is given by

$$d(R, Q) = |z_0 - z_1|.$$

Now, the distance we are looking for is the length of the line segment PQ, which is the hypotenuse of the right triangle PQR (see Figure 1.4).

This is a right triangle, so we can obtain the length using the Pythagorean Theorem. So, we get

$$d(P, Q) = \sqrt{d(P, R)^2 + d(R, Q)^2}.$$

Substituting in what we found above, we obtain

$$d(P, Q) = \sqrt{(x_0 - x_1)^2 + (y_0 - y_1)^2 + (z_0 - z_1)^2}.$$

This completes the proof.   ■

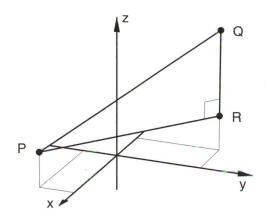

FIGURE 1.4

4. *"Looking back."* What we have presented is our version of the proof. You may find that you need to include more details. By all means, go ahead. If you had to stop and say, "where did that come from?" make sure you answer yourself. Write it in the text (you aren't going to sell this book back anyway, right?), or keep a notebook of "proofs with commentary." Note that though we used pictures to illustrate the ideas in our argument, a picture will not, in general, substitute for a proof. However, it can really clarify an idea. Don't rely on a picture, but don't be afraid to use one either.   ○

Solutions to Exercises

Solution to Exercise (1.1).
We are given that this code was created through a shift of the alphabet. Thus once we determine one letter, the other letters are easily found. Since we have a one-letter word, we'll start with it. Thus "W" must represent either the letter "I" or the letter "A." Checking both shifts of the alphabet

$(W \rightarrow I, X \rightarrow J, Y \rightarrow K, Z \rightarrow L, A \rightarrow M, B \rightarrow N, C \rightarrow O$, etc.

$$W \rightarrow A, X \rightarrow B, Y \rightarrow C, Z \rightarrow D, A \rightarrow E, B \rightarrow F, C \rightarrow G, \text{ etc.})$$

we find that if "W" represents the letter "I", then "E" must represent the letter "Q." The fact that "EO" appears as a word and "O" would represent the letter "A" in our coded text implies that "EO" would be the word "QA," which is an interesting combination of letters, but hardly a word. Thus, "W" cannot represent the letter "I" and therefore "W" represents "A."

Using the shift described above and replacing the corresponding letters, we find that the code says the following.

"THIS ENCODING ALGORITHM IS CALLED A CAESAR CIPHER. IT IS VERY EASY TO BREAK, RIGHT?"

In fact, the Caesar cipher is quite easy to break. If this interests you, a very readable history of coding theory is presented by S. Singh in *The Code Book*, [78].

Spotlight: George Pólya

György Pólya (1887–1985), referred to as George Pólya in his later years, was born and raised in Hungary. He studied in Vienna and in Budapest, where he received his doctorate in 1912. One of his influential teachers was Leopold Fejér. In his book [67, p. 39], Pólya refers to Fejér as "an inspiring teacher who had a great deal of influence on Hungarian mathematicians of the time." The two primary places that Pólya taught were the Eidgenössische Technische Hochschule (ETH) in Zürich, Switzerland and Stanford University in Palo Alto, California.

Though Pólya's mother tongue was Hungarian, he worked in the Swiss-German speaking part of Switzerland and he spoke French with his wife from Neuchâtel, a city in the French speaking part of Switzerland. In school he also learned Latin and Greek. (See [67, p. 11].) Pólya later emigrated to the United States where he taught and lectured in English. He published mathematical papers in Hungarian, German, French, English, Italian, and Danish.

Pólya contributed to original research in probability, geometry, number theory, real and complex analysis, graph theory, combinatorics, and mathematical physics. His name is connected to many mathematical ideas and constructions. To name just a few of his achievements, we mention that in probability there is a Pólya distribution and he is credited with introducing the idea and the term of "random walk." But Pólya was not only recognized as an excellent scholar of mathematics, he was also an excellent teacher of mathematics. His heuristic approach to problem solving is outlined in *How to Solve It.* This book had a profound influence on the teaching of mathematics. It has sold over one million copies and is translated into over 20 languages. Records kept by the ETH in Zürich show that Pólya was the advisor of 14 thesis students there and, according to [62], he was the advisor of 9 more students at Stanford.

The Mathematical Association of America (MAA) gives an annual Pólya award. According to the MAA website, "This award, established in 1976, is named after the renowned teacher and writer, and is given for articles of expository excellence published in the *College Mathematics Journal.*"

To learn more about George Pólya and his approach to problem solving, we recommend reading his book *How to Solve It*, [66], the picture book [67] (which contains a short biography), or consulting the more in-depth account of Pólya's life, written by his former student at Stanford, [3]. The article [85] is based on interviews with Pólya and appeared in an issue of *Mathematics Magazine* entirely devoted to Pólya and his work.

Problems

Problem 1.1.
Here is a problem intended to help you work through "the list." After this, you are on your own.

Find a word (written in standard capital letters) that is unchanged when reflected in a horizontal line and in a vertical line. The word must appear in a dictionary (in a language of your choice) in order to be a valid solution.

1. *"Understanding the problem."* We need to find a word. We are given information about the letters that make up this word. There are two conditions: Two different reflections should not alter the word.

   Try these two reflections on a word, say on SOLUTION, to make sure you understand the problem.

2. *"Devising a plan."* We have to find the connection between what we are given and what we have to find.

   Which letters of the alphabet satisfy each of the two conditions? Both conditions?

   Find a word that is not changed if it is reflected in a horizontal line.

   Find a word that is not changed if it is reflected in a vertical line. Formulate the exact conditions for this exercise; that is, state the letters that can be used and how they must be arranged.

3. *"Carrying out the plan."* Find a word that satisfies the conditions given above.

4. *"Looking Back."* Are there other solutions?

Problem 1.2.

Find a word (written in standard capital letters) that reads the same forward and backward and is still the same forward and backward when rotated around its center 180°. Your solution needs to appear in a standard dictionary of some language.

Problem 1.3.

Solve the following anagrams. The first three are places (in the geographical sense), and the fourth is a place you might live in. All can be rearranged to form a single word.

   (a) NOVA CURVE;

   (b) NINE SLAP NAVY;

   (c) I HELD A HIP PAL;

   (d) DIRTY ROOM.

   Note: You may have to find out exactly what an anagram is. This is part of Pólya's first point on the list.

Problem 1.4.

Suppose n teams play in a single game elimination tournament. How many games are played?

An example of such tournaments are the various categories of the U. S. Open tennis tournament; for example, women's singles.

Note: Pay special attention to the first entry of Pólya's list: "Is it possible to satisfy the condition?"

Problem 1.5.

Suppose you are all alone in a strange house. There are seven identical closed doors. The bathroom is behind exactly one of them. Is it more likely, less likely, or equally likely that you find the bathroom on the first try than on the third try? Why?

Problem 1.6.

The following message is encoded using a shifted alphabet just as in Exercise 1.1. (Of course, the shift number n is not the same as in the exercise!) What does the message say?

RDSXCVIWTDGNXHUJCLTLXAAATPGCBDGTPQDJIXIAPITG

Problem 1.7.

Give a detailed description of all points in three-space that are equidistant from the x-axis and the yz-plane. Once you decide on the answer, write the solution up carefully. Pay particular attention to your notation.

Problem 1.8.

The following is a classic problem in mathematics. Though there are many variations of this problem, the standard one is the following.

You are given 12 coins that appear to be identical. However, one of the coins is counterfeit, and the weight of this coin is slightly different than that of the other 11. Using only a two-pan balance, what is the smallest number of weighings you would need to find the counterfeit coin? (Think about a simpler, similar problem.)

(See I. Peterson's web site [64] for a discussion of this problem.)

Problem 1.9.

Let n be an odd integer. Prove that $n^3 - n$ is divisible by 24.

The following two problems are only appropriate if you took at least two semesters of calculus. Though you may have worked these before, the idea is to work them again paying close attention to the final presentation. Make sure you define all variables. Use complete sentences, with proper punctuation.

Problem 1.10.
Find the volume of a spherical cap if the height is 2 m and the radius of the rim of the cap is 5 m.

Problem 1.11.
We have two circular right cylinders of radius 1 each. The axes of the two cylinders intersect at a right angle. Find the volume of the solid that both cylinders have in common.

Tips on Doing Homework

Your instructor will probably ask you to work many of the exercises and problems in this text. If there is one thing mathematicians agree on, it is that you learn mathematics by doing it. Here are some tips on how to get started.

- Make sure you know what the rules are. Some instructors do not want you to get help from someone else. Other instructors encourage working together in groups. Ask, if you are not clear about the policy.
- If you are permitted to work together, form a study group. A small group of two to four people usually works best. Get together on a regular basis and discuss the assigned problems.
- Read the questions carefully. If there is a term that you do not know, look it up.
- Before you get started, read over the text and the notes from class, paying particular attention to definitions, theorems, and previous exercises. It isn't unusual to spend several hours on a single problem at this point. Doing mathematics means pondering a problem for hours, days, weeks, even years (though we have tried not to

pose problems that will take you years to solve). Working two hours on one problem, thinking about it as you go through your day and then spending another two hours on it the next day is fairly common practice for students at this level.

- Once you have read over the text, looked over the relevant definitions, worked through the examples and tried to solve the problem, you will be well on your way towards understanding the problem. If you can't get started, at least you will know which questions to ask. Seek help from your instructor or other students (if your instructor allows this).

- Once you have a solution to a problem, look at it critically. Check that it is correct. Put it down. Come back to it later. Do you still understand everything? Is it still correct? (As you can imagine, this is very important.) Can you simplify it? If you work with someone else have them read it over. *Never hand in your first draft of a solution to a problem.*

- Writing a solution means convincing a reader that the result is correct. There can be *no* gaps or errors. Explain each step—don't assume that the reader knows what you are thinking. Keep a reader in mind as you write, and remember that the instructor or anyone else who already knows the solution is *not* really your target audience. Though that may be the person for whom the solution is intended, it is your job to convince the reader that each step in your solution is correct. Perhaps a better audience to keep in mind is someone who knows the material from the class, but not the solution to the problem.

- Write up your final solution very carefully and neatly. The reader shouldn't find him or herself proving things for you—you should do that for him or her. Staple pages together so that the reader may have the pleasure of reading your entire proof in the correct order and its entirety.

2

CHAPTER

Logically Speaking

Suppose your friend tells you that Mr. Hamburger is German or Swiss. You happen to know that Mr. Hamburger is not Swiss. Using your powers of reasoning, you decide that Mr. Hamburger is German. Note that this argument can be generalized, because it doesn't really depend on Mr. Hamburger being Swiss or German. If your friend said that "A or B is true" and you happened to know that "B is not true," you would conclude that "A is true." This is an example of a valid argument. Now suppose your friend tells you that Mr. French eats only pickles on Wednesday, and only chocolate on Monday. You know that Mr. French is eating chocolate that day. Now what can you say? While you may conclude that Mr. French has odd eating habits, you would not have used a logically valid argument to do so. In this example, there is really only one thing you can conclude. We'll return to this at the end of this chapter.

In order to understand an argument, we must be able to read and comprehend the sentences that compose it. We need to be able to tell whether the sentences in our argument are true or false, and whether they follow logically from the previous ones. So now for a definition. A **statement** is a sentence that is either true or false (but not both). "Two is not a prime number" is an example of a (false) statement. "Do you love me?" is not a statement. Below are some examples and some nonexamples of statements. These will be your first examples of nonexamples.

Exercise 2.1.
Which of the sentences below are statements and which are not?
  (a) It is raining outside.
  (b) The professor of this class is a woman.
  (c) Two plus two is five.
  (d) $X + 6 = 0$.
  (e) Seven is a prime number.
  (f) All odd numbers are prime.
  (g) This sentence is false.   ○

   Because English usage and mathematical usage may differ slightly, we must be certain that we understand our statements before we construct arguments. We now carefully study the truth or falsity of statements. Our treatment is brief. (See [56] for a more detailed study of mathematical logic.)
   The rules of logic that we present in this chapter should work for all statements, and not just particular ones. For this reason, we introduce letters such as P, Q, R, or S to represent statements. Thus P will have two possible truth values: true, denoted T, or false, denoted F. We can negate P or combine it with Q by saying things like:

> Not P.
> P and Q .
> P or Q.
> If P, then Q.
> P if and only if Q.

Such symbolic sentences will be called **statement forms**. A precise definition of statement form will be given once we have precise definitions of the **connectives** "not," "and," "or," "if ..., then ...," and "if and only if."
   In the English language we might say

> It's raining.
> It is not raining.
> If it is raining, the sky is grey.

It is raining or it is snowing.
It is cold and it is snowing.
It is snowing if and only if it is cold.

Let's start with the simplest case. Suppose your teacher says, "This book has a blue cover." Taking a quick glance at the cover, you can decide on the truth value of that statement; namely that it is false. In order to have a true statement, you could say, "This book does not have a blue cover." If we have a statement form P, the **negation of** P is the statement form "not P." Under what circumstances should the negation of P be true or false? We will always use the notation $\neg P$ for "not P." If P is true, then $\neg P$ should be false. If P is false, then $\neg P$ should be true. We can summarize all the possibilities in a truth table as follows:

P	**¬P**
T	F
F	T

What about combining two statement forms, P and Q, into one statement form as "P or Q"? In this sentence, it is particularly important to distinguish between mathematical usage of the word "or" and everyday speech. For example, if we say, "You can have cake or ice cream," it could be that you can have both. If we say, "The door is open or closed," it cannot be that the door is both open and closed. English statements involving the word "or" are often ambiguous; in mathematics, ambiguity is generally frowned upon. The statement form "P or Q" is called a **disjunction** and is denoted $P \vee Q$. In mathematics, a disjunction is true when P alone is true, Q alone is true, or both P and Q are true. So in mathematics, you can always have your cake and ice cream.

Exercise 2.2.
Complete the truth table for $P \vee Q$.

P	**Q**	**P ∨ Q**
T	T	
T	F	
F	T	
F	F	

○

The statement form "*P* and *Q*" is called a **conjunction** and is denoted $P \wedge Q$. We will have you fill in the truth table for "*P* and *Q*" below. It should be clear that this will be true when both *P* and *Q* are true, and false otherwise.

Exercise 2.3.

Complete this truth table.

P	**Q**	**P ∧ Q**
T	*T*	
T	*F*	
F	*T*	
F	*F*	

○

Now consider the statement form "If *P*, then *Q*." This statement form is called an **implication** and is often stated as "*P* implies *Q*" and written $P \rightarrow Q$. (Note that though English usage of the word "implies" may suggest a relationship between *P* and *Q*, our analysis of truth values has assumed no connection at all between *P* and *Q*.) There are equivalent ways of stating an implication, and some will require careful thinking on the reader's part. Remember as you read on that "If *P*, then *Q*" may also be stated as

Q if *P*.
P is sufficient for *Q* (meaning *P* is enough to make *Q* happen).
Q is necessary for *P* (if *P* happened, then *Q* must have happened).
P only if *Q* (same as above; if *P* happened, then *Q* must have happened).
Q whenever *P*.

The statement form *P* in each of these formulations is called the **antecedent**, and *Q* is called the **conclusion**. Under what conditions is an implication true? false? Let's begin with an example you are all familiar with. Suppose we say to our son,

"If you clean your room, then you can go to Henry's house."

Under what conditions would he feel that we had lied? In the example, the antecedent, *P*, is "you clean your room." The conclusion,

Q, is "you can go to Henry's house." Well, if our son cleans his room and we let him go to Henry's, everybody is happy. That implication should be true. So if P is true and Q is true, the whole statement should be true. Also, it should be as clear to you as it will be to our son, that if he cleans his room and we do not let him go to Henry's, we lied. So, if P is true, and Q is false, the implication should be false. Now what if he doesn't clean his room? We never discussed this possibility. So no matter what we decide here, we have not lied. In this situation, the statement is not false; hence we consider it to be true. So if P is false, no matter what the truth value is of the conclusion, we will consider the implication to be true.

Summarizing this discussion, the only way that the implication "If P, then Q" can be false is if P is true and Q is false. In the exercise below you will sum up this discussion in the form of a truth table.

Exercise 2.4.
Complete this truth table.

P	Q	P → Q
T	T	
T	F	
F	T	
F	F	

○

It is often helpful to rephrase a statement, making sure that you maintain the same true and false values. The statement form "*P if and only if Q*" is called an **equivalence**, and we will write this as $P \leftrightarrow Q$. This is the same statement form as "*(P only if Q) and (P if Q)*." In view of the discussion above, we see that this is also $(P \rightarrow Q) \wedge (Q \rightarrow P)$. Thus the truth table for the equivalence is

P	Q	P → Q	Q → P	P ↔ Q
T	T	T	T	T
T	F	F	T	F
F	T	T	F	F
F	F	T	T	T

Look down the final column and you'll see that the equivalence is true precisely when P and Q are both true or both false.

The statement form "If P, then Q" is also written as $P \Rightarrow Q$, and "P if and only if Q" might be written as $P \Leftrightarrow Q$ or "P iff Q."

Having now studied the connectives, we are ready for our definition of a statement form. A **statement form** is a letter representing an unspecified statement or an expression built from such letters using connectives.

Now consider the two statement forms $\neg(P \vee Q)$ and $\neg P \wedge \neg Q$. In the next exercise, you will find the truth table for each of these expressions and compare them.

Exercise 2.5.
Write out the truth tables for $\neg(P \vee Q)$, $\neg P \wedge \neg Q$, and $(\neg(P \vee Q)) \leftrightarrow (\neg P \wedge \neg Q)$. What can you conclude? ○

A statement form for which the final column in the truth table consists of all T's is called a **tautology**. A statement form for which the final column is all F's is called a **contradiction**. Two *statement forms*, P and Q, are said to be **(logically) equivalent** if $P \leftrightarrow Q$ is a tautology, and two *statements* are **equivalent** if they can be obtained from two equivalent statement forms by consistently replacing the letters by English statements.

In view of Exercise 2.5, we see that $\neg(P \vee Q)$ and $\neg P \wedge \neg Q$ are equivalent statement forms. Thus the statement "It is not the case that Rachel or Leah won the race" is equivalent to "Rachel did not win the race and Leah did not win the race." (Why?)

While it is very important to be able to restate something in an equivalent form, it is equally important that you be able to negate a statement. Some useful negations appear in the exercises and problems. The negation of an implication is particularly important in mathematics. If you think about integers and the sentence "If x is prime, then x is odd or $x = 2$," you can see that even a relatively simple implication might be difficult to negate. Let's begin with something simpler.

Exercise 2.6.
Construct the truth table for $P \to Q$, and the truth table for $\neg P \vee Q$. What do you notice? Now construct a truth table for $(P \to Q) \leftrightarrow$

$(\neg P \vee Q)$. What conclusion can you make? Finally, find an equivalent way to write $\neg(P \to Q)$. ○

If all went well, you noticed that $P \to Q$ is equivalent to $\neg P \vee Q$, and therefore the negation of "If P, then Q" is "P and not Q." Let's return to

$$\text{"If } \underbrace{x \text{ is prime}}_{P} \text{ , then } \underbrace{x \text{ is odd or } x = 2}_{Q} \text{."}$$

Negating this leads to

$$\text{"} \underbrace{x \text{ is prime}}_{P} \text{ and it is } \underbrace{\text{not the case that } x \text{ is odd or } x = 2}_{\neg Q} \text{."}$$

While this is the negation, it isn't really as helpful as it might be. So we now negate the disjunction "x is odd or $x = 2$" and combine it with our previous work to obtain

$$\text{"} x \text{ is prime and } x \text{ is not odd and } x \neq 2 \text{."}$$

Refining this further, we would probably say something like "x is prime, even and not equal to two." The negation of an implication is something you should learn well now because it arises frequently. Here are some examples for you to try.

Exercise 2.7.
Negate the following. It's interesting to note that you can negate a statement even if you don't understand what it says. It is easier to get it right, though, if you understand the statement.
(a) If I go to the party, then he is there.
(b) If x is even, then x is divisible by 2.
(c) If a function is differentiable, then it is continuous.
(d) If x is a natural number, then x is even or x is odd. ○

Exercise 2.8.
Which of the following are equivalent to each other? All the answers have appeared in this chapter.

$$P \to Q, \neg(P \vee Q), \neg(P \wedge Q), P \wedge \neg Q, \neg(P \to Q),$$
$$P \vee \neg Q, \neg P \vee \neg Q, \neg P \wedge \neg Q, \neg P \vee Q.$$

So let's apply what we have learned in this chapter to Mr. French, who eats only pickles on Wednesday and only chocolate on Monday. One statement is that "if it is Wednesday, then Mr. French eats only pickles." We let W represent the statement "it is Wednesday," and P the statement "Mr. French eats only pickles." Thus, we know that $W \to P$ is true. (If you thought we should have said $W \wedge P$ is true, note that we do not know that the statement W is true, so we must use the implication here.) The second is "if it is Monday, then Mr. French eats only chocolate." Letting M denote "it is Monday" and C the statement that "Mr. French eats only chocolate" we may write what we are given as $M \to C$. Finally we are told that "Mr. French is eating chocolate." From this we can conclude that $\neg P$ is true. Let's put this together.

1. $W \to P$,
2. $M \to C$, and
3. $\neg P$.

Now, it's fairly clear that the second statement is irrelevant. So let us look at the truth tables for the first and third statements (for convenience, we combine the two tables):

W	P	W → P	¬P
T	T	T	F
T	F	F	T
F	T	T	F
F	**F**	**T**	**T**

We know that both $W \to P$ and $\neg P$ are true, and from our truth table we see that there is only one time that this happens: when both W and P are false. So there you have it. All we can conclude is that it is not Wednesday.

People differ in their approaches to problems. In the example above, you might have found it easier not to rewrite the problem. That's fine. On the other hand, when a problem starts to confuse you, looking at it as we have here will often help you figure out how to attack a problem.

Solutions to Exercises

Solution to Exercise (2.2).
The truth table for $P \vee Q$ is

P	Q	P $\vee$ Q
T	T	T
T	F	T
F	T	T
F	F	F

Solution to Exercise (2.3).
The truth table for $P \wedge Q$ is

P	Q	P $\wedge$ Q
T	T	T
T	F	F
F	T	F
F	F	F

Solution to Exercise (2.4).
The truth table for $P \rightarrow Q$ is

P	Q	P $\rightarrow$ Q
T	T	T
T	F	F
F	T	T
F	F	T

This is the same as the truth table for $\neg P \vee Q$.

Solution to Exercise (2.6).
In the solution to Exercise 2.4, we noted that $P \rightarrow Q$ and $\neg P \vee Q$ are equivalent. Thus $\neg(P \rightarrow Q)$ is equivalent to $\neg(\neg P \vee Q)$, which is, as we have seen in Exercise 2.5, equivalent to $P \wedge \neg Q$. In words, the negation of "If P, then Q" is "P and not Q."

Solution to Exercise (2.7).
More than one answer is possible but they must be equivalent, of course.
(a) I go to the party and he is not there.
(b) One answer is: x is even and x is not divisible by 2.
(c) A function is differentiable and it is not continuous.
(d) One answer is: x is a natural number and x is not even and x is not odd. Equivalently, we could say: x is a natural number, and x is neither even nor odd.

Problems

Problem 2.1.
In the following implications, identify the antecedent and the conclusion.
(a) If it is raining, I will stay home.
(b) I wake up if the baby cries.
(c) I wake up only if the fire alarm goes off.
(d) If x is odd, then x is prime.
(e) The number x is prime only if x is odd.
(f) You can come to the party only if you have an invitation.
(g) Whenever the bell rings, I leave the house.

Problem♮ 2.2.
Construct a truth table for $\neg(\neg P)$. Is this what you expect? Why?

Problem♮ 2.3.
Find a statement form, S, equivalent to $\neg(P \vee Q)$ and show that it is logically equivalent by constructing the truth table for "S if and only if $\neg(P \vee Q)$" and showing that this statement form is a tautology.

Problem 2.4.
Write out the truth table for the statement form $P \rightarrow \neg(Q \wedge \neg P)$. Is this statement form a tautology, a contradiction, or neither?

Problem 2.5.
Write out the truth table for the statement form $(P \rightarrow (\neg R \vee Q)) \wedge R$. Is this statement form a tautology, a contradiction, or neither?

Problem 2.6.
Negate the sentences below and express the answer in a sentence that is as simple as possible.
   (a) I will do my homework and I will pass this class.
   (b) Seven is an integer and seven is even.
   (c) If T is continuous, then T is bounded.
   (d) I can eat dinner or go to the show.
   (e) If x is odd, then x is prime.
   (f) The number x is prime only if x is odd.
   (g) If I am not home, then Sam will answer the phone and he will tell you how to reach me.
   (h) If the stars are green or the white horse is shining, then the world is eleven feet wide.

Problem 2.7.
For each of the cases below, write a tautology using the given statement form. For example, if you are given $P \vee \neg Q$ you might write $(P \vee \neg Q) \leftrightarrow (Q \rightarrow P)$.
   (a) $\neg(\neg P)$;
   (b) $\neg(P \vee Q)$;
   (c) $\neg(P \wedge Q)$;
   (d) $P \rightarrow Q$.

Problem 2.8.
When we write, we should make certain that we say what we mean. If we write $P \wedge Q \vee R$, you may be confused, since we haven't said what to do when you are given a conjunction followed by a disjunction. Put parentheses in to create a statement form with the given truth table.

P	**Q**	**R**	**P $\wedge$ Q $\vee$ R**
T	T	T	T
T	T	F	T
T	F	T	T
T	F	F	F
F	T	T	T
F	T	F	F
F	F	T	T
F	F	F	F

Problem 2.9.
For each of the cases below, write a contradiction using the given statement form. For example, if you are given $\neg(\neg P)$ you might write $\neg(\neg P) \leftrightarrow \neg P$.
  (a) $P \to Q$;
  (b) $\neg(P \vee Q)$;
  (c) $\neg P \vee \neg Q$;
  (d) $P \leftrightarrow Q$.

Problem 2.10.
Consider the statement "It snows or it is not sunny."
  (a) Find a different statement that is equivalent to the given one.
  (b) Find a different statement that is equivalent to the negation of the given one.

Problem 2.11.
The following problem is well known. Many different versions of this problem appear in [80].
      On a certain island, each inhabitant is either a truth-teller or a liar (and not both, of course). A truth-teller always tells the truth and a liar always lies. Arnie and Barnie live on the island.

(a) Suppose Arnie says, "If I am a truth-teller, then each person living on this island is either a truth-teller or a liar." Can you say whether Arnie is a truth-teller or liar? If so, which one is he?

(b) Suppose that Arnie had said, "If I am a truth-teller, then so is Barnie." Can you tell what Arnie and Barnie are? If so, what are they?

3

CHAPTER

Introducing the Contrapositive and Converse

In the last chapter we saw that two statement forms, P and Q, that have the same truth table are equivalent. This was also expressed by showing that the equivalence, $P \leftrightarrow Q$, is a tautology. When you are confronted with a mathematical statement that you need to prove, you will often find it helpful to paraphrase it. You will use tautologies to do so, since you don't want to change the truth value of your statement. Some useful tautologies appear below and throughout this chapter.

Theorem 3.1.
Let $P, Q,$ and R denote statement forms. Then the following are tautologies:

(DeMorgan's laws)	$\neg(P \vee Q) \leftrightarrow (\neg P \wedge \neg Q)$;
	$\neg(P \wedge Q) \leftrightarrow (\neg P \vee \neg Q)$;
(Distributive property)	$(P \wedge (Q \vee R)) \leftrightarrow ((P \wedge Q) \vee (P \wedge R))$;
	$(P \vee (Q \wedge R)) \leftrightarrow ((P \vee Q) \wedge (P \vee R))$;
(Double negation)	$\neg(\neg P) \leftrightarrow P$;

31

(Associative property) $(P \wedge (Q \wedge R)) \leftrightarrow ((P \wedge Q) \wedge R);$
$(P \vee (Q \vee R)) \leftrightarrow ((P \vee Q) \vee R);$

(Commutative property) $(P \wedge Q) \leftrightarrow (Q \wedge P);$
$(P \vee Q) \leftrightarrow (Q \vee P).$

At this point, you should be able to construct the truth tables for everything above and you should be able to show that all of them are tautologies.

Exercise 3.2.
Negate the following:
(a) $(P \wedge Q) \vee (P \wedge R);$
(b) $P \rightarrow (Q \wedge R).$ ○

Tautologies allow us to replace one statement by another. For example, suppose you want to show that an integer is odd or prime. You can show that the integer is prime or odd; that won't change things because these two statements are equivalent. This is a fairly obvious change that usually won't make much of a difference. The same holds if you want to show x is prime and odd; you can show that it is odd and prime if that's easier and you will have accomplished the same thing. Similarly, if you want to show that it is not the case that x is prime and odd, you can show that x is not prime or not odd.

For implications, restating what you want to prove can really make a difference. We need to make sure, however, that what we have is equivalent to our original statement. So recall that we showed, in the last chapter, that $P \rightarrow Q$ is equivalent to $\neg P \vee Q$.

Now consider $\neg Q \rightarrow \neg P$, which is called the **contrapositive** of the implication $P \rightarrow Q$. We need to compare the two truth tables below:

P	Q	P → Q
T	T	T
T	F	F
F	T	T
F	F	T

P	Q	¬Q → ¬P
T	T	T
T	F	F
F	T	T
F	F	T

So the fact that the truth tables are the same tells us that the statement forms are logically equivalent. What this means to us is that, if we are trying to prove that an implication is true and we don't see how to do it, we should consider the contrapositive of that statement. Here's how it works in practice.

Theorem 3.3.
Let x be an integer. If x^2 is odd, then x is odd.

First we need to understand the problem. What does it mean for a number x to be odd? It means that there is an integer n such that $x = 2n + 1$. So we are assuming that $x^2 = 2n + 1$ for some integer n and trying to show $x = 2m + 1$ for some integer m. It's hard to see where to go from here, we think.

Remember that Pólya suggests restating the problem, so let's try that. Let P be the sentence "x^2 is odd" and Q be the sentence "x is odd." Then we see that we wish to prove that $P \rightarrow Q$ is true. But this is logically equivalent to $\neg Q \rightarrow \neg P$, which translates into "If x is not odd, then x^2 is not odd." We can do better than that, since an integer is either odd or even. So we can show that "If x is even, then x^2 is even" and that will be equivalent. Let's see if that's easier.

Theorem (Contrapositive of the statement of Theorem 3.3).
Let x be an integer. If x is even, then x^2 is even.

The first step is to understand the problem. The second step is to prove it. We'll do that here:

"*Understanding the problem.*" When is an integer even? When it is of the form $x = 2n$, where n is an integer. So we need to show that $x^2 = 2m$, where m is an integer, assuming that $x = 2n$, where n is an integer. We began by understanding the problem, now we are ready to solve it.

Proof.
Let x be even. Then there is an integer n such that $x = 2n$. Therefore, $x^2 = (2n)^2 = 2(2n^2)$. Let $m = 2n^2$. Then $x^2 = 2m$ and m is an integer. Therefore x^2 is even. ∎

Of course the original theorem is now also proven since it is equivalent to the one we proved. Thus, using the contrapositive is one possible way to attempt to prove that an implication is true. We will soon have a number of ways to attack a problem. Try to keep them all in mind.

Some other related remarks: Notation is more important than it may seem. In the theorem above, we assume that x is even and try to show x^2 is even. If we assume that $x = 2n$ and accidentally try to show $x^2 = 2n$ (rather than $x^2 = 2m$), we're stuck because we erroneously assumed that $x = x^2$. In other words, our notation would force us to show that $x = 0$ or $x = 1$, which is not what we should be doing. We introduced an error because of poor notation. So it's important that one symbol be an n and one be an m.

Also, note that we begin the proof by saying what we are assuming, and end the proof by saying what we are concluding. That helps the reader too. Finally, we keep checking that m and n are integers. That's because that is very important; if they weren't integers, x wouldn't have to be even.

So the contrapositive was very helpful here. You do need to be careful though. It must be the contrapositive and not the converse. The **converse** of an implication $P \to Q$ is the statement form $Q \to P$. Looking at the truth tables for each of these given below,

P	Q	P → Q
T	T	T
T	F	F
F	T	T
F	F	T

and

P	Q	Q → P
T	T	T
T	F	T
F	T	F
F	F	T

,

we see that they are different. Unfortunately, though the contrapositive and converse of a statement are really very different, students often confuse them. We'll take just a moment to convince you that it is very important not to do this.

Suppose our statement is, "If I am a Hobbit, then I am under 5 feet tall." This is a true statement, as every Tolkien reader knows. The converse is "If I am under 5 feet tall, then I am a Hobbit." This latter statement is not true, since lots of children are under 5 feet tall, but most of them are not Hobbits. As a mathematical example, consider the sentence about integers "If x is seven, then x is prime," and its converse "If x is prime, then x is seven." Recall that an integer p is **prime** if $p > 1$ and p cannot be written as a product of two positive integers, both different from p. Thus, the original sentence is true for all x, while the converse above is not. On the other hand, you agree that for all x the contrapositive "If x is not prime, then x is not seven," is true, as it must be. But this is trickier when we don't really understand what we are saying as well as we understand this statement. Remember, make sure you understand the problem.

Exercise 3.4.
Consider the sentence "If n is odd, then $n^2 - n - 6$ is even."
  (a) State the contrapositive.
  (b) State the converse.   ○

Solutions to Exercises

Solution to Exercise (3.2).
The equivalences are given below.
  (a) The negation may be stated as $(\neg P \vee \neg Q) \wedge (\neg P \vee \neg R)$, since

$$\neg((P \wedge Q) \vee (P \wedge R)) \leftrightarrow (\neg(P \wedge Q) \wedge \neg(P \wedge R))$$
$$\leftrightarrow ((\neg P \vee \neg Q) \wedge (\neg P \vee \neg R)).$$

  (b) The negation may be stated as $P \wedge (\neg Q \vee \neg R)$, since

$$\neg(P \rightarrow (Q \wedge R)) \leftrightarrow (P \wedge \neg(Q \wedge R))$$
$$\leftrightarrow (P \wedge (\neg Q \vee \neg R)).$$

Solution to Exercise (3.4).
  (a) The contrapositive is "If $n^2 - n - 6$ is odd, then n is even."
  (b) The converse is "If $n^2 - n - 6$ is even, then n is odd."

Problems

Problem♭ 3.1.
(a) Let x be an integer. Prove that if x is odd, then x^2 is odd. Make sure you state your assumption as the first line and your conclusion as the last line.
(b) State the contrapositive of what you just proved.
(c) Combining the result of part (a) with Theorem 3.3 gives a stronger result. Say precisely what that result is.

Problem 3.2.
For each of the following, write out the contrapositive and the converse of the sentence.
(a) If you are the President of the United States, then you live in a white house.
(b) If you are going to bake a soufflé, then you need eggs.
(c) If x is a real number, then x is an integer.
(d) If x is a real number, then $x^2 < 0$.

Problem 3.3.
State the contrapositive of each of the following.
(a) If it rains, then it pours.
(b) If I had a bell, I would ring the bell in the morning.
(c) The house is red, if the house is not blue.
(d) Dinner is cooked only if I make it.

Problem 3.4.
State the converse of each of the following.
(a) If it rains, then it pours.
(b) If I am young, then I am restless.
(c) I am alone if it is Saturday.
(d) I eat fish only if it is cooked.

Problem 3.5.
Let x and y be real numbers. Show that if $x \neq y$, then $2x + 4 \neq 2y + 4$. (Hint: Use the contrapositive.)

Problem 3.6.

Matilda always eats at least one of the following for breakfast: cereal, bread, or yogurt. On Monday, she is especially picky.

If she eats cereal and bread, she also eats yogurt. If she eats bread or yogurt, she also eats cereal. She never eats both cereal and yogurt. She always eats bread or cereal.

Can you say what Matilda eats on Monday? If so, what does she eat?

Problem 3.7.

Consider the following statement.

> If the coat is green, then the moon is full or the cow jumps over it.

(a) This unusual statement is composed of several substatements. Identify each substatement, give it a letter, and write down the original statement using these letters and logical connectives.

(b) Using the symbols introduced in (a), find the contrapositive of the original statement. Rewrite the contrapositive as an English sentence.

(c) Find the converse of the *original* statement, writing the sentence and its converse in symbols, and then rewriting the converse in words.

(d) Find the negation of the *original* statement, writing the sentence and its converse in symbols, and then rewriting the converse in words.

(e) Are some of the statements in this problem (either the original or the ones you obtained) equivalent? If so, which ones?

Problem 3.8.

Consider the two statement forms $P \rightarrow Q$ and $P \rightarrow (Q \vee \neg P)$.

(a) Make a truth table for each of these statement forms.

(b) What can you conclude from your solution to part (a)?

Problem 3.9.

Karl's favorite brownie recipe uses semisweet chocolate, very little flour, and less than 1/4 cup sugar. He has four recipes: one French, one Swiss, one German, and one American. Each of the four has at least two of the qualities Karl wants in a brownie recipe. Exactly three use very little flour, exactly three use semisweet chocolate, and exactly three use less than 1/4 cup sugar.

The Swiss and the German recipes use different kinds of chocolate. The American and the German recipes use the same amount of flour, but different kinds of chocolate. The French and the American recipes use the same amount of flour. The German and American recipes do not both use less than 1/4 cup sugar.

Karl is very excited because one of these is his favorite recipe. Which one is it?

Problem 3.10.
Let n be an integer. Prove that if $3n$ is odd, then n is odd.

Problem 3.11.
Prove that if x is odd, then $\sqrt{2x}$ is not an integer.

Problem 3.12.
Let x and y be real numbers. Show that if $x \neq y$ and $x, y \geq 0$, then $x^2 \neq y^2$.

4

CHAPTER

Set Notation and Quantifiers

Before we get to the heart of this chapter, it will be useful to have notation for the things we frequently work with. A **set** is a collection of objects. The objects in the set are called **elements** or **members** of the set. We will write $x \in X$ to indicate that x is an element of X. (Some people read $x \in X$ as "x belongs to X," others read it "x is an element of X.") Usually we will be considering things of a particular type. The set of all possible objects that are considered in the context in which we work is called the **universe**. We will usually denote it by X. In some cases the universe may consist of all real numbers, or it may consist of all right triangles; it might even consist of all cows living in France. The set may consist of all positive real numbers, all isosceles right triangles, or all white cows living in France. And the elements might be the real number π, the isosceles right triangle with legs of length 1, or Farmer Boursin's white cow Elsie, who lives in Dijon, France.

When it is implicitly clear what the universe is, we may not mention it specifically. But when there is any doubt at all, we will carefully state what the universe is. Once we do that, we can denote a set by writing $S = \{x \in X : x \text{ satisfies } P\}$. The brackets indicate that we are talking about a collection of objects, called elements;

$x \in X$ tells us where these elements live, and P is a property these elements have.

In this class, as well as others, some sets show up a lot and we have special notation for them. Notation should always be chosen carefully, as these have been. Most mathematicians agree on these, so don't make up your own notation and make sure you recognize what these are when they are used:

The **natural numbers** $\mathbb{N} = \{0, 1, 2, 3, \ldots\}$.
The **integers** $\mathbb{Z} = \{\ldots, -2, -1, 0, 1, 2, \ldots\}$.
The **positive integers** $\mathbb{Z}^+ = \{1, 2, 3, \ldots\}$.
The **real numbers** $\mathbb{R}$.
The **plane** $\mathbb{R}^2 = \{(x, y) : x, y \in \mathbb{R}\}$.
For $n \in \mathbb{Z}^+$, **Euclidean n-space** $\mathbb{R}^n = \{(x_1, x_2, \ldots, x_n) : x_j \in \mathbb{R} \text{ for } j = 1, 2, \ldots, n\}$.
The **rational numbers** $\mathbb{Q} = \{p/q : p, q \in \mathbb{Z} \text{ and } q \neq 0\}$.
The **complex numbers** $\mathbb{C} = \{a + bi : i^2 = -1 \text{ and } a, b \in \mathbb{R}\}$.

Some authors include zero in $\mathbb{N}$ and others don't. If you look in another text, make sure you know what convention they follow.

For real numbers a and b with $a \leq b$, the set $[a, b] = \{x \in \mathbb{R} : a \leq x \leq b\}$ is called the **closed interval** from a to b. The sets $[a, \infty) = \{x \in \mathbb{R} : a \leq x\}$ and $(-\infty, b] = \{x \in \mathbb{R} : x \leq b\}$ are called unbounded closed intervals. For $a < b$, the set $(a, b) = \{x \in \mathbb{R} : a < x < b\}$ is called the **open interval** from a to b. We shall see that (a, b) can be interpreted several ways, and you should be able to decide which from the way it is used. You've done this in the past. For example, you have certainly had courses where (x, y) denotes a point and, in the same course, (x, y) might denote an open interval. Unbounded open intervals are defined, with appropriate changes, as we defined unbounded closed intervals.

Exercise 4.1.
Find a (different) useful way to describe the following sets (your useful way could be a sketch):
  (a) $\{x \in \mathbb{Z} : x^2 = 1\}$;
  (b) $\{x \in \mathbb{N} : x^2 = 1\}$;
  (c) $\{(x, y) \in \mathbb{R}^2 : y = 0\}$;
  (d) $\{(x, y, z) \in \mathbb{R}^3 : z = 0\}$;

(e) $\{x \in \mathbb{Z} : x \text{ is even}\}$;
(f) $\{(m, n) : m, n \in \mathbb{Z}\}$.   ○

Now we can talk about slightly more complicated sentences. Think of the difference between the statements "In every box there is a prize" and "In some box there is a prize." Obviously, if you had to choose (and if it were the same prize) you would go with the first one. In mathematics, in order to determine the truth or falsity of a statement, we need to know whether we are talking about a particular x or all x. What we mean should be clear from the context. Letters like x that stand for elements of the universe are called **variables**. The phrases "for all," "for every," "for some," or "there exists," quantify variables. "For all," or ∀, is the **universal quantifier** and "there exists," or ∃, is the **existential quantifier**.

After agreeing that the universe consists of all real numbers, consider the following statement: "For all x it is the case that $x^2 - 1 \leq 0$." We know that we are asking that for every x, something must happen. It just so happens that this statement is false, but it is still a clear statement. For all x is usually written ∀x. So we could write

$$\forall x, x^2 - 1 \leq 0.$$

What follows the words "For all x" in our statement is another sentence that we could denote by p, but since p is a sentence involving x we write $p(x)$. The statement above is of the form

$$\forall x, p(x).$$

One more remark about the example above. Suppose the universe is (still) the real numbers, but we want to make this a statement about positive integers only. In that case, we can express our statement symbolically as follows:

$$\forall x, (x \in \mathbb{Z}^+ \rightarrow (x^2 - 1 \leq 0)).$$

For a different example, suppose that our universe is the set of integers and consider the sentence, "There is an integer x such that $x = 0$." This, too, is a statement, and happens to be true. This statement can be expressed symbolically by

$$\exists x, (x = 0)$$

and is read as "there exists x such that $x = 0$." This statement is of the form

$$\exists x, p(x).$$

One more remark about the last example. If we had chosen the set of the real numbers as the universe, we would express our statement symbolically as

$$\exists x, (x \in \mathbb{Z} \wedge x = 0).$$

This becomes very important when you are negating statements. You can easily see why, too: if you negate $x \in \mathbb{Z}$ and $\mathbb{Z}$ is your universe, then there are no x left, but if you negate $x \in \mathbb{Z}$ and $\mathbb{R}$ is your universe, there are still plenty of x left to worry about. So make sure that you give careful consideration to your universe before beginning a problem.

Exercise 4.2.
Write the statements below in symbols, assuming that the universe is $\mathbb{R}$ throughout. Make sure that you clearly quantify x; is it "all x" or "some x"?
   (a) For all x, it is the case that x is an integer.
   (b) There exists an integer x such that $x > 0$.
   (c) There is a rational number x such that $x^2 + 1 = 0$.
   (d) For every real number x, there exists a real number y such that $x < y$.
   (e) There is a real number y such that $x < y$ for all x.
   (f) If x is a real number, then $x^2 + 1 \neq 0$.
   (g) A real number x satisfies $x^2 > 0$, if $x \neq 0$.
   (h) If $x > 0$, then $x > 4$ or $x < 6$.   ○

We negated conjunctions, disjunctions, and implications. Now we will think about the negation of a quantified statement.
   Suppose we have the statement "Every cow is black." How would we negate it? One pretty useless way is to say "Not every cow is black." It's better to say "Some cow is not black." So a useful negation of

$$\forall x, p(x)$$

is

$$\exists x, \neg p(x).$$

Similarly, if we say, "There exists a black cow" a useful negation is "No cow is black." So a negation of

$$\exists x, p(x)$$

is

$$\forall x, \neg p(x).$$

You will find that sometimes you can negate a sentence directly and other times you need to convert to symbols. Here is another example.

Example 4.3.
Negate the sentence "People who live in glass houses should not throw stones."

We will assume that the universe is the set of all people. What does this say? First, it says something about all people who live in glass houses. So we will use the quantifier "for all" and x will denote a person. The notation $g(x)$ will mean that x lives in a glass house. The notation $t(x)$ will mean that x should throw stones. So our sentence becomes "For all x, if $g(x)$, then $\neg t(x)$." If you can negate it now, go ahead. If not, go through the steps below. You should provide reasons why each step below is correct:

1. $\neg(\forall x, (g(x) \to \neg t(x)))$;
2. $\exists x, \neg(g(x) \to \neg t(x))$;
3. $\exists x, \neg(\neg g(x) \vee \neg t(x))$;
4. $\exists x, (g(x) \wedge t(x))$.

The last sentence says that the negation of "People who live in glass houses should not throw stones" is "There exists a person who lives in a glass house and should throw stones." ○

We emphasize that while it is good to practice these symbolic manipulations, it is also important to understand what you are doing. Sometimes you will find it easier to use the symbolic notation and sometimes you won't. Make sure you keep in mind what the sentence says, and whether or not your answer seems reasonable. Before you go off on your own, we'll do a fairly complicated example together.

Example 4.4.
Suppose our universe is the set of real numbers and we wish to negate the statement "For every rational number x, there exists an integer n that is greater than x."

So let's try it. First we note that "For every rational number x" means that we are being told that "if x is a rational number" something will happen. What? There will exist an integer bigger than x. So this is an implication of the form "For all x, if x is a rational number, then there exists an n such that n is an integer and $n > x$." Sometimes it is easier to understand a statement if we replace the various subsentences with symbolic representations. We use

$p(x)$ for x is a rational number,
$q(n)$ for n is an integer, and
$r(n, x)$ for $n > x$.

Using this notation, we have

$$\forall x, (p(x) \rightarrow \exists n, (q(n) \wedge r(n, x))).$$

Let's try to negate this quantified statement form one step at a time, starting from the outside.

We know that when we negate "for all" it becomes "there exists." In other words, we can replace $\neg(\forall x, \cdots)$ with $\exists x, \neg(\cdots)$. So, here's where we are now:

$$\neg(\forall x, (p(x) \rightarrow \exists n, (q(n) \wedge r(n, x))))$$

is equivalent to

$$\exists x, \neg(p(x) \rightarrow \exists n, (q(n) \wedge r(n, x))).$$

Now we negate the implication. From the last chapter we know that $\neg(P \rightarrow Q)$ is equivalent to $P \wedge \neg Q$. We're up to

$$\exists x, (p(x) \wedge \neg(\exists n, (q(n) \wedge r(n, x)))).$$

So the only thing left to do is negate Q, which is the expression $\exists n, (q(n) \wedge r(n, x))$. At least this is simpler than what we started with! Now $\exists$ will change to $\forall$ and so we need only worry about $q(n) \wedge r(n, x)$. But that's a conjunction. So the final step is to negate that, and we know the negation of the conjunction will become $\neg q(n) \vee \neg r(n, x)$. So here's where we are now:

$$\exists x, (p(x) \wedge (\forall n, (\neg q(n) \vee \neg r(n, x)))).$$

We've done what we were asked to do, in a sense, but our answer is still in symbols. Let's translate back:

"There exists an x such that x is a rational number and for all n, either n is not an integer or n is not greater than x."

And finally (you should explain how we get the following),

"There is a rational number x such that for all n, if n is an integer, then $n \leq x$." ○

Not all negations are this complicated, but even in simpler statements there are things you should be wary of. Consider the two statements about real numbers: $\forall x, \exists y, x + y = 0$ and $\exists y, \forall x, x + y = 0$. Assuming the universe is the set of real numbers, what's the difference between these two statements? In the first, we say that for each x we can find a y with $x + y = 0$. That's a statement you have known to be true for years, ever since you learned about $-x$. On the other hand, the second statement says that there exists a y such that for all x, we have $x + y = 0$. That statement is false, because the same y would have to work for all x. What's the moral of this story? That the order of the quantifiers is very important.

Exercise 4.5.
Negate the statements (a)–(h) of Exercise 4.2. ○

Solutions to Exercises

Solution to Exercise (4.1).
There are many possible answers. We list some below:
  (a) $\{1, -1\}$;
  (b) $\{1\}$;
  (c) the x-axis in $\mathbb{R}^2$;
  (d) the xy-plane in $\mathbb{R}^3$;
  (e) $\{2n : n \in \mathbb{Z}\} = \{\ldots, -2, 0, 2, \ldots\}$;
  (f) the set of all points in $\mathbb{R}^2$ such that both the x and y coordinates are integers.

Solution to Exercise (4.2).
Note that the universe was assumed to be $\mathbb{R}$.
  (a) $\forall x, x \in \mathbb{Z}$.
  (b) $\exists x, ((x \in \mathbb{Z}) \wedge (x > 0))$.
  (c) $\exists x, ((x \in \mathbb{Q}) \wedge (x^2 + 1 = 0))$.
  (d) $\forall x, \exists y, (x < y)$.
  (e) $\exists y, \forall x, (x < y)$.
  (f) $\forall x, \neg(x^2 + 1 = 0)$.
  (g) $\forall x, (\neg(x = 0) \rightarrow x^2 > 0)$.
  (h) $\forall x, (x > 0 \rightarrow ((x > 4) \vee (x < 6)))$.

Solution to Exercise (4.5).
Note that the universe was assumed to be $\mathbb{R}$.
  (a) There exists an x such that x is not an integer.
  (b) For all x, either x is not an integer or x is nonpositive (or both). This is equivalent to: For all x, if x is an integer, then x is nonpositive.
  (c) For all x, if x is a rational number, then $x^2 + 1 \neq 0$.
  (d) There exists an x such that for all y we have $x \geq y$.
  (e) For all y, there exists an x such that $x \geq y$.
  (f) For some x, it is the case that $x^2 + 1 = 0$.
  (g) For some x, we have $x \neq 0$ and $x^2 \leq 0$.
  (h) There exists a positive real number x such that $x \leq 4$ and $x \geq 6$.

Problems

Tips on Quantification on page 51 summarizes many of the major points in this chapter. You may find it helpful to read these tips before working the problems below.

Problem 4.1.
Write the following statements symbolically.
- (a) For every x, there is a y such that $x = 2y$.
- (b) For every y, there is an x such that $x = 2y$.
- (c) For every x and for every y, it is the case that $x = 2y$.
- (d) There exists an x such that for some y the equality $x = 2y$ holds.
- (e) There exists an x and a y such that $x = 2y$.

Problem 4.2.
Which of the statements in Problem 4.1 are true if the universe for both x and y is the set of the real numbers?

Problem 4.3.
Which of the statements in Problem 4.1 are true if the universe for x is the set of the real numbers and the universe for y is the set of the integers?

Problem 4.4.
Negate the statements in Problem 4.1.

Problem 4.5.
Negate the following sentences. If you don't know how to negate it, change it to symbols and then negate. State the universe whenever it is not evident.
- (a) For all $x \in \mathbb{R}$, we have $x^2 > 0$.
- (b) Every odd integer is nonzero.
- (c) If I am hungry, then I eat chocolate.
- (d) For every girl there is a boy she doesn't like.
- (e) There exists x such that $g(x) > 0$.
- (f) For every x there is a y such that $xy = 1$.
- (g) There is a y such that $xy = 0$ for every x.
- (h) If $x \neq 0$, then there exists y such that $xy = 1$.

(i) If $x > 0$, then $xy^2 \geq 0$ for all y.
(j) For all $\epsilon > 0$, there exists $\delta > 0$ such that if x is a real number with $|x - 1| < \delta$, then $|x^2 - 1| < \epsilon$.
(k) For all real numbers M, there exists a real number N such that $|f(n)| > M$ for all $n > N$.

Problem 4.6.
Consider the following statement.

> For all positive integers x, there exists a real number y such that for all real numbers z, either $y = z^x$ or $z = y^x$.

(a) Write this statement using symbols and appropriate quantification. Use $\mathbb{R}$ for the universe of all variables.
(b) Once you have written this statement in symbols, negate the (symbolic) statement that you obtained.

Problem 4.7.
Consider the following statement:

$$\forall x, ((x \in \mathbb{Z} \wedge \neg(\exists y, (y \in \mathbb{Z} \wedge x = 7y))) \rightarrow (\exists z, (z \in \mathbb{Z} \wedge x = 2z))).$$

(a) Negate this statement.
(b) Write the original statement as an English sentence.
(c) Which statement is true, the original one or the negation? Explain your answer.

Problem 4.8.
Write each of the statements below using symbolic notation. In this problem, use $\mathbb{R}$ as the universe for all variables involved.
(a) There is an integer that is bigger than its square.
(b) Every rational number is the product of two irrational numbers. (Note: A real number x is irrational if $x \notin \mathbb{Q}$.)
(c) There are integers m and n such that for each rational number x, either $m < nx$ or $n < mx$.
(d) Every rational number is the solution of an equation $ax+b = 0$, where a and b are integers.

Problem 4.9.
Why is this joke supposed to be funny? A physicist, chemist, and a mathematician are traveling through Switzerland. From the train they spot a cow grazing in the field. The chemist gazes out the window and says, "Ah, all the cows in Switzerland are brown." The physicist says, "No, no. You can't conclude that. You can only say that some of the cows in Switzerland are brown." The mathematician says, "No, no, no. All you can say is that there is a cow in Switzerland that is brown on one side."

Problem 4.10.
For each of the following, state the converse, the contrapositive, and the negation of each (the negation of the statement, the converse, and the contrapositive). State the universe, if appropriate and quantify anything that is quantifiable.
  (a) Madeleine waters the plants only if it is Tuesday.
  (b) If I ski, I will fall.
  (c) Windows break if you throw balls through them.
  (d) If I negate a sentence, then I always do it wrong.
  (e) I will come only if you invite me.
  (f) For all positive real numbers x, there exists an integer n such that $1/n < x$.
  (g) If x is a nonzero real number, then $x^2 \neq 0$.
  (h) If x is a nonzero real number, then there exists a real number y such that $x \cdot y = 1$.
  (i) If x and y are even integers, then $x + y$ is an even integer.

Problem 4.11.
Find a different useful description of the following:
  (a) $\{x \in \mathbb{R} : x^2 = 2\}$;
  (b) $\{(x, y) \in \mathbb{R}^2 : x = y\}$;
  (c) $\{x \in \mathbb{N} : x \leq 0\}$;
  (d) $\{x \in \mathbb{Z} : x^2 > 0\}$.

Problem 4.12.
Write each of the following in set notation.
  (a) The set of all odd integers.
  (b) The set of all points in the xy-plane above the line $y = x$.

(c) The set of all points in the xy-plane that are inside the circle of radius one.

(d) The set of all irrational numbers.

Problem 4.13.

Decide whether sentence (3) is true if sentences (1) and (2) are both true. Give reasons for your answers.

(a) (1) Everyone who loves Bill loves Sam.
   (2) I don't love Sam.
   (3) I don't love Bill.

(b) (1) If Susie goes to the ball in the red dress, I will stay home.
   (2) Susie went to the ball in the green dress.
   (3) I did not stay home.

(c) (1) If l is a positive real number, then there exists a real number m such that $m > l$.
   (2) Every real number m is less than t.
   (3) The real number t is not positive.

(d) (1) Every little breeze seems to whisper Louise or my name is Igor.
   (2) My name is Stewart.
   (3) Every little breeze seems to whisper Louise.

(e) (1) There is a house on every street such that if that house is blue, the one next to it is black.
   (2) There is no blue house on my street.
   (3) There is no black house on my street.

(f) Let x and y be real numbers.
   (1) If $x > 5$, then $y < 1/5$.
   (2) We know $y = 1$.
   (3) So $x \leq 5$.

(g) Let M and n be real numbers.
   (1) If $n > M$, then $n^2 > M^2$.
   (2) We know $n < M$.
   (3) So $n^2 \leq M^2$.

(h) Let x, y, and z be real numbers.
   (1) If $y > x$ and $y > 0$, then $y > z$.
   (2) We know that $y \leq z$.
   (3) Then $y \leq x$ or $y \leq 0$.

Tips on Quantification

- Check the universe for each of the variables. Write it down, if it is not self-evident.
- Suppose a statement restricts the variable x to a proper subset A of the universe as in the statement form, "For all $x \in A$, property $p(x)$ holds." Since x is universally quantified, this is an implication of the form

$$\forall x, (x \in A \rightarrow p(x)).$$

- Suppose a statement restricts the variable x to a proper subset A of the universe as in the statement form, "For some $x \in A$, property $p(x)$ holds." Since x is existentially quantified, this is a conjunction of the form

$$\exists x, (x \in A \wedge p(x)).$$

- Simple statements are usually easy to negate. Just do it.
- Complicated statements will often resist a "just do." Write them out in symbols first. Make sure you know what the quantifier is on every variable. Check for the various ways one can say "if..., then..."
- Do not use logical connectives ($\neg, \wedge, \vee, \rightarrow, \leftrightarrow$) between quantifiers. (Do not write "$\forall x \vee \forall y \cdots$" or "$\forall x \wedge \forall y \cdots$.")
- Know the rules. You must know how to negate existential quantifiers, universal quantifiers, conjunctions, disjunctions, and implications. The most important negation is also the one students frequently forget: the negation of an implication.
- Practice: Every time you get a definition or theorem, try negating it. If you can't, this might indicate that you do not fully understand it.

If you think you need more practice, here it is. In what follows, unless otherwise stated, all variables are real numbers, and ϵ and δ represent positive real numbers. Negate all of these.

(a) For every ϵ, there exists δ such that $\delta < \epsilon$.
(b) Let $a \in \mathbb{R}$. For every ϵ there exists δ such that for every $x \in \mathbb{R}$, if $|x - a| < \delta$, then $|x^2 - a^2| < \epsilon$.
(c) Let $x \in \mathbb{R}$. Then $x < x + \epsilon$ for all $\epsilon > 0$.

(d) For every integer n, there exists $x > n$ such that $x^2 > n^2$.

(e) For every $\epsilon > 0$ there exists an integer N such that $1/n < \epsilon$ for all $n \geq N$.

(f) For all x, either $x < 0$ or $x > 0$.

(g) For all x, there exists an integer n such that $n > x$.

(h) For all x, y, and z, if $x < y$ and $z < 0$, then $zx > zy$.

(i) Let x and y be real numbers. If $x < y + \epsilon$ for all $\epsilon > 0$, then $x \leq y$.

5

CHAPTER

Proof
Techniques

In this chapter, we introduce you to some of the most common proof techniques. The three methods we will examine in this section are:

- direct proof (just get started and keep going),
- proof by contradiction (show that the negation of the statement you wish to prove implies the impossible), and
- proof in cases (which may be used when conditions dictate that different situations occur).

There are many more. For example, another proof technique that you may be familiar with from the study of calculus is the method of exhaustion, such as computing area or volume calculations by "filling up the object" with a sequence of more familiar smaller sets. Sometimes these techniques are used in combination. Some other methods, such as proof of existence and uniqueness of an object or proof using the contrapositive of the statement, will appear in subsequent chapters.

The first example is a direct proof. We want to show that "If A, then B is true." So we do it in our most direct manner: We start with A and keep going until we get to B. Before getting started, we make sure we know the meaning of every word in the implication and we try to make sure that the implication is true.

Theorem 5.1.

If a, b, and c are integers such that a divides b and a divides c, then a divides $b + c$.

"*Understanding the problem.*" Okay, before we get started, let's identify the hypothesis and conclusion. What are they? The hypothesis is a, b, and c are integers such that a divides b and a divides c. We get to start with that. What does a divides b mean? Well, we don't know yet, so let's think about that. It would mean that when we divide b by a we get an integer. So this would mean $b = an$, where $n \in \mathbb{Z}$. So we say a divides b if and only if there is an integer n such that $b = an$. Since we have already defined everything here, we understand the problem and we feel confident—raring to go, in fact. What's the conclusion we need to come to? The conclusion is a divides $b + c$, and we know what this means because we understand "divides."

"*Devising a plan.*" So we know that, in the notation we used above, $b = am$ and $c = an$ where m and n are both integers. We need to show that a divides $b + c$, or that there is an integer j with $b + c = aj$. Looking at what we were given and what the desired conclusion is should suggest the plan.

Proof.

Since a, b, and c are integers such that a divides b and a divides c, we know that there exist integers m and n such that $b = am$ and $c = an$. Therefore, $b + c = am + an = a(m + n)$. Since $m + n$ is an integer, a divides $b + c$. ∎

"*Looking back.*" Let's admire this proof for a minute. It's so lovely. There are complete sentences, periods, and all symbols are carefully defined. We say where we are starting; that is, what the assumption is, and we end by saying what the conclusion is. Just in case the reader hasn't noticed, though, we indicate that we are done by adding the little box, ∎. Other people use Q.E.D. *(quod erat demonstrandum* which is Latin for *which was to be demonstrated).* Your proofs should be just as appealing as the one above.

What follows is an example of a proof by contradiction, sometimes referred to as *reductio ad absurdum.* The idea of such a proof is that we suppose that what we wish to conclude is false and show that something really silly happens (hence the absurdum). Below is an example of this idea that goes back to the Pythagoreans. This is one of two proofs presented by G. H. Hardy in his famous book *A Mathematician's Apology* [35], as an example of a beautiful proof. (The first proof in Hardy's text is in the problems. If you haven't read his book, it is another one that we highly recommend.)

Theorem 5.2.
The number $\sqrt{2}$ is not rational.

"*Understanding the problem.*" Before we begin, we make sure that we know what all the words mean, what we are assuming, and what we are trying to prove. A rational number is a number of the form p/q where p and q are integers, and q is nonzero. So we need to show that $\sqrt{2}$ is not of this form; that is, there are no integers p and q (with q nonzero) such that $\sqrt{2} = p/q$. That may seem like a tall order, since it seems to mean we have to look through all possible integers! This leads directly to:

"*Devising a plan.*" Perhaps it would be easiest to assume $\sqrt{2} = p/q$ (with p and q integers and $q \neq 0$) and see what, if anything, happens. This is precisely the idea behind proof by contradiction.

Proof.
Suppose, to the contrary, that $\sqrt{2}$ is rational. Then there exist integers p and q (with q nonzero) such that $\sqrt{2} = p/q$. We may assume that p and q have no common factor, for if they did, we would simplify and begin again. Now, we have that $\sqrt{2}q = p$. Squaring both sides, we obtain $2q^2 = p^2$. Thus p^2 is even. Since p^2 is even, we know from Problem 3.1 that p must be even. Therefore, $p = 2m$ for some integer m. This means that $2q^2 = 4m^2$. Dividing, we see that $q^2 = 2m^2$. But this means that q^2 is even. Again we know from Problem 3.1 that q is even. So p and q have a common factor 2, which is completely absurd, since we assumed they had no common factor. Therefore our assumption that $\sqrt{2}$ is rational must be wrong and we have completed the proof of the theorem. ∎

"*Looking back*." Note that we slipped in a reference to Problem 3.1. If we hadn't, you would have read "Since p^2 is even, p must be even." Your reaction to this could have been "Oh yeah, we did that already." That's fine. But you could also have stopped, tried to think about why it is true, tried to prove it, and so on. That's fine too, in some sense, but you don't want to re-prove everything we have already done. So if the writer tells the reader why something is true, it saves the reader valuable time. Or, you could also have skipped right over it, never worrying about why it is true. That's not fine. You need to understand each sentence in a proof!

Knowing how to split a proof into cases, which we will refer to as a "proof in cases," is something that will be extremely useful too. Here is an example of something defined in cases. Once we understand this definition, we'll prove something using it.

For a real number x, the **absolute value** of x is defined in cases by

$$|x| = \begin{cases} x & \text{if } x \geq 0 \\ -x & \text{if } x < 0 \end{cases} .$$

Is this what you were expecting the definition to be? If not, let's make sure it agrees with what you were expecting. If $x = 3$, then $x \geq 0$, and we conclude that $|3| = 3$. If $x = -3$, then $x < 0$, and we conclude that $|-3| = -(-3) = 3$. If you feel comfortable with this definition, you are ready to move on to the theorem. If not, work out a few more examples and then move on.

Theorem 5.3.
Let x and y be real numbers. Then $|xy| = |x||y|$.

We made sure that we understood the definition of absolute value before proceeding to the theorem, so we understand the problem. Let's think about devising a plan.

"*Devising a plan.*" Absolute value was defined in cases, and therefore $|xy|$ depends on whether $xy \geq 0$ or $xy < 0$. The first, $xy \geq 0$, is actually two cases again: $xy > 0$ or $xy = 0$. What are the possibilities? Well, $xy > 0$ would mean that both $x > 0$ and $y > 0$, or both $x < 0$ and $y < 0$. The case $xy = 0$ would mean that $x = 0$ or $y = 0$. The final possibility, $xy < 0$, would mean that one of the two, x or

y, is negative and the other is positive. It seems that we have four cases to consider: both x and y positive, both negative, at least one of the numbers is zero and one of the two numbers negative while the other is positive.

Proof.

First, suppose that $x > 0$ and $y > 0$. Then $xy > 0$ and we have $|xy| = xy$, $|x| = x$, and $|y| = y$. Therefore,

$$|xy| = xy = |x||y|,$$

and we have established the result in this case.

Second, suppose that $x < 0$ and $y < 0$. Then $xy > 0$ and we have $|xy| = xy$, $|x| = -x$, and $|y| = -y$. Therefore,

$$|xy| = xy = (-x)(-y) = |x||y|,$$

and we have the result for this case as well.

Third, suppose that either $x = 0$ or $y = 0$. Then $xy = 0$ and we have $|xy| = 0$, and either $|x| = 0$ or $|y| = 0$. Therefore,

$$|xy| = 0 = |x||y|,$$

establishing the result in this case too.

For our final case, suppose that one number is positive and the other is negative. Thus, we may assume that $x < 0$ and $y > 0$. Then $xy < 0$ and we have $|xy| = -(xy)$, $|x| = -x$, and $|y| = y$. Therefore,

$$|xy| = -(xy) = (-x)y = |x||y|.$$

We have now established the result for all four possible cases and we may conclude that $|xy| = |x||y|$ for all real numbers x and y. ∎

Once again, look at the form of the proof. There are four cases and we tell the reader which case we are discussing before we discuss it. We can conclude something in each case, but it isn't until we cover all four possible cases that we can write "we may conclude that $|xy| = |x||y|$ for all real numbers x and y."

It will also be helpful to know how to show something is not true. A statement whose truth is anticipated, but for which we have no proof yet is called a **conjecture**. There are many different ways that one might arrive at a conjecture. It can be due to the intuition or insight of a great mathematician, or it can be a generalization of observations gleaned from many examples. The latter has become more common in recent years, in part due to the capabilities of powerful calculators and computers. Once we find a proof, the conjecture turns into a theorem. The most famous example in recent history is a proof by Andrew Wiles. In 1995, Wiles turned Fermat's last conjecture into Fermat's last theorem, [86]. (Watch the excellent Nova episode "The Proof" for the full story on the history of Fermat's last theorem, [7].)

It's important to note, however, that just because you believe something might be true, doesn't mean that it necessarily is true. Sometimes you will find that a conjecture someone else has made (or even one that you have made) is, in fact, false. In these cases, you need to find an example of something that satisfies the hypotheses of your conjecture, but not the conclusion. An example is the following conjecture of Pierre de Fermat—one of the very few of his conjectures that turned out to be wrong.

Consider numbers of the form $2^{2^m} + 1$, where m is a natural number. The first number, $2^{2^0} + 1 = 3$ is prime. The second, $2^{2^1} + 1 = 5$ is also prime, as are the third, fourth, and fifth numbers. In fact, Fermat conjectured that if m is a nonnegative integer, then $2^{2^m} + 1$ is prime. In 1732, the Swiss mathematician, Leonhard Euler, showed that this was false by showing that the sixth number in this list, $2^{2^5} + 1 = 4294967297$ can be factored. In fact, our calculator tells us that $2^{2^5} + 1 = 641 \cdot 6700417$. Thus Fermat's conjecture is false.

An example that shows that a statement is false is called a **counterexample.** You only need one to show something is false!

Problems

Problem 5.1.
Below is the other proof Hardy chose to present ([35, pp. 92–94]). This theorem and its proof were known to Euclid, and appear in the *Elements* IX 20, [38]. Can you read and understand this proof? Read the whole thing. Underline anything you don't understand the first time. Reread it slower this time. Underline anything you can't figure

out. You may need to spend 10 minutes on each sentence; you may not. Then write the general idea of the proof in "street talk." A bright, interested twelve year old should be able to follow your outline of the proof.

Before you begin, make sure you understand what will be assumed and what we will try to do. Make sure you know what all the words mean. "Infinite" has not yet been defined; prime number has.

Theorem 5.4.
There are infinitely many prime numbers.

Proof.
To prove this statement suppose, to the contrary, that there are finitely many primes. Then we may write these finitely many primes in ascending order as

$$2, 3, 5, \ldots, N,$$

where N is the largest prime. Now consider the number M defined by

$$M = (2 \cdot 3 \cdot 5 \cdots \cdot N) + 1.$$

If M is prime, then M is a prime that is larger than the largest prime N. Therefore, we must conclude that M is not prime, and so it is divisible by some prime number, P. However, P must appear in the list of primes

$$2, 3, 5, \ldots, N,$$

which we gave earlier. But when we divide M by P, we obtain a remainder of 1. Therefore, P cannot be a factor of M, and we have contradicted our assumption that there are finitely many primes. Thus, there exist infinitely many primes. ■

Problem 5.2.
Prove that if n is an integer, then $4n^2 + 4n + 8$ is an even integer. What kind of proof did you use?

Problem 5.3.
Prove that if n is an integer, then $n^2 + 3n + 2$ is an even integer. What method of proof did you use?

Problem 5.4.

Provide counterexamples to each of the following.

(a) Every odd number is prime.

(b) Every prime number is odd.

(c) For every real number x, we have $x^2 > 0$.

(d) For every real number $x \neq 0$, we have $1/x > 0$.

(e) Every function $f : \mathbb{R} \to \mathbb{R}$ is linear (of the form $mx + b$).

Problem 5.5.

Define two sets, A and B, by

$$A = \{x \in \mathbb{Z} : x = 2n \text{ for some } n \in \mathbb{Z}\} \text{ and}$$
$$B = \{x \in \mathbb{Z} : x = 2m + 1 \text{ for some } m \in \mathbb{Z}\}.$$

(a) Using these definitions, give a rigorous proof that A and B have no elements in common. Make sure you write out all details.

(b) What type of proof did you use in part (a)?

Problem 5.6.

Let n be an integer. Prove that if n^2 is divisible by 3, then n is divisible by 3.

Problem 5.7.

Show that $\sqrt{3}$ is not rational. (You may want to use the result of Problem 5.6 to work this problem.)

Problem 5.8.

Prove that $\sin^2 x \leq |\sin x|$ for all $x \in \mathbb{R}$.

Problem♭ 5.9.

Let x be a real number.

(a) Prove that $-|x| \leq x \leq |x|$.

(b) Let $a \geq 0$. Prove that $|x| \leq a$ if and only if $-a \leq x \leq a$.

(c) Prove the theorem below.

Theorem 5.5 (The triangle inequality).

Let x and y be real numbers. Then $|x + y| \leq |x| + |y|$.

Problem 5.10.

Prove **the lower triangle inequality**: Let x and y be real numbers. Then

$$||x| - |y|| \le |x - y|.$$

Problem 5.11.

Find all points in the xy-plane that lie on the surface

$$4 = 5(x - 3)^2 + 3(y - \pi)^2 + 2(z + 2)^2.$$

Write up your solution carefully. What method of proof did you use?

Problem 5.12.

Let n be an integer. Prove that if $n^2 - (n - 2)^2$ is not divisible by 8, then n is even.

Problem 5.13.

Prove that if $p(x) = a_n x^n + a_{n-1} x^{n-1} + \cdots + a_0$, where $a_0, \ldots, a_n \in \mathbb{R}$, is a polynomial, then p can have at most n roots. (Some remarks are in order here. To work this problem, you must understand it. Recall that $c \in \mathbb{R}$ is a root of a polynomial p if $p(c) = 0$. In order to restate the problem, you also need to recall that if c is a root of p, then $x - c$ is a factor of p.)

Problem 5.14.

Consider the following statement.

$$\forall x, (x \in \mathbb{Z}^+ \to \exists y, \exists z,$$
$$((y \in \mathbb{Q}) \wedge (z \in \mathbb{Q}) \wedge (yz \ne 0) \wedge (x^2 = y^2 + z^2))).$$

(a) Change this symbolic statement to an English sentence.
(b) Prove the statement you found in (a).

Tips on Definitions

In your previous courses, you may or may not have had to memorize definitions. Now it becomes essential that you memorize them, understand them, and investigate them before venturing on to use them. Here are some suggestions on how to do these things.

- The first step is to make sure you know the definition. This does not mean that you highlight it with a marker and read it over a few times. It means that you, first of all, understand it, and, second of all, memorize it. You must know whether the quantifiers are "for all" or "there exist," you must know what order they come in, you must watch the order on implications, and you must be sure that what you write is correct. Every single itty bitty detail must be correct or chances are that your definition is wrong.
- It's very difficult to memorize something you don't understand. So once you see a definition (in bold black print in this book) write it down and think about what it means.
- Give many examples, until you feel that you know what an example looks like.
- Negate the definition and try to find nonexamples (that show when things won't satisfy the definition).
- Go back and see if you can write out the definition without looking at it. Wait a few hours and do that again. If anything is out of place, ask yourself if it matters. If it does, repeat the appropriate steps here.
- Definitions are often stated as implications. This leads students to ask if the definition is an equivalence. The answer is "yes." Consider the following definition: "An integer m is even if there exists an integer n such that $m = 2n$." Since this is how we defined "even," we also mean that "if m is even, then there exists an integer n such that $m = 2n$."

Some teachers and students find it helpful to make definition notebooks. In such a notebook, you will do all the steps above as often as necessary. We heartily recommend such an approach.

6
CHAPTER

Sets

Recall from Chapter 4 that a set S is a collection of objects. The objects that make up the set are called the elements or members of the set. A set has a defining property, and it is used to determine whether or not an element belongs to the set: To decide whether or not x is in the set S, you need to see whether x satisfies this defining property p. The **empty set** is the set with no elements, and is denoted by $\emptyset$.

Once we have the defining property, there are often several ways to describe a set. If there aren't too many elements in the set, then we can list all elements: $B = \{Benny, Betty, Billy, Bobby\}$. If the elements come from a well-known larger set X and satisfy a defining property p, we may write $\{x \in X : p(x)\}$. This is read "the set of all elements of X satisfying property p." We may think of X as the universe in this context.

Note that a set is described by its elements—not by the order we put the elements in the set, or whether we put an element in more than once. Thus the set $\{1, 2, 3\}$ is the same as the set $\{1, 1, 3, 2\}$.

Exercise 6.1.
For each of the following sets, say what the universe is and write out the defining property. For example, if we wish to describe the

set of all women, the universe might be all people, and the defining property would be "x is a woman." Use complete sentences.
  (a) The collection A of all members of the school band.
  (b) The collection B of all irrational numbers.
  (c) The collection of all prime numbers greater than or equal to 4 and less than 7. ○

Exercise 6.2.
Care needs to be used when creating a defining property. What is wrong with each of the following?
  (a) The collection C of all pretty people in Luxembourg.
  (b) The collection D of all collections that do not contain themselves as an element. ○

   The notation we have described so far in this chapter is not the only acceptable notation. For example, if we know what our universe is, there may be no reason to repeat it in the notation. Therefore, we may write $\{x \in X : p(x)\}$, or we may simply write $\{x : p(x)\}$. The next exercise introduces you to a slightly different way of describing a set.

Exercise 6.3.
Let $S = \{x \in \mathbb{Z} : x = 2n + 1$ for some $n \in \mathbb{Z}\}$ and $T = \{s^2 : s \in S\}$. The notation for T is different from the notation we have discussed thus far in the chapter, yet you can still determine T. Write out a description of T using the same notation as the one used for S. Then write out a description of S using the same notation as the one used for T. ○

Exercise 6.4.
Consider the set A of nonzero integers.
  (a) Write this set using the notation $A = \{x \in S : p(x)\}$.
Use what you learned in previous chapters to answer the following questions.
   Define a new "multiplication" on A by $x \star y = 2xy$ for $x, y \in A$. For parts (b) and (c) below, either prove the statement or give a

counterexample to it. (If you find you cannot answer the questions below, read the discussion following part (c).)

(b) If $x, y \in A$, then $x \star y \in A$.

(c) There exists an element $y \in A$ such that $x \star y = x$ for every $x \in A$.

If you really can't get started, then you probably didn't understand the problem. One way to begin is to pick numbers for x and y and try them out until you get a feel for this new multiplication. Once you understand it, try rewriting the statements so that they make sense to you. For example, in (b), replace the conclusion $x \star y$ by its definition to obtain "If $x, y \in A$, then $2xy \in A$." All this should help. Remember, the most important thing is to get started. ○

A set A is a **subset** of a set B or, equivalently, A is **contained** in B, if every element of A is an element of B. We will write $A \subseteq B$ to indicate that A is a subset of B. This is depicted in Figure 6.1.

Notice that A is always a subset of itself: $A \subseteq A$. However, a subset can also be truly smaller, and we often find it necessary to use our notation to emphasize this. We say that A is a **proper subset** of B if $A \subseteq B$ and $A \neq B$, and we will write $A \subset B$.

Showing that a set A is contained in another set B turns out to be one of the most important tasks in mathematics. One way to show that a set A is contained in a set B is to do exactly what the definition says; take an arbitrary element of the set A and then show that this element is in set B.

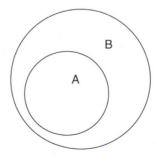

FIGURE 6.1 $A \subseteq B$

Example 6.5.
In Exercise 6.3 we used two sets, S and T, where $S = \{2n+1 : n \in \mathbb{Z}\}$ and $T = \{s^2 : s \in S\}$. Show that $T \subseteq S$. Is T a proper subset of S?

Remember, to prove set inclusion we have to take an arbitrary element in the set T and then show that this element is in the set S. So, for our proof of containment, we will begin with $x \in T$, and attempt to end our proof with $x \in S$.

As we will see in future chapters, we can often devise a plan for a proof of this type by writing out what we know ($x \in T$) at the top of the page, and what we want to show ($x \in S$) at the bottom. You have probably attempted proving things this way before: you work from the top down, and from the bottom up. So our plan might look like

$$x \in T,$$

large space

$$x \in S.$$

But $x \in T$ means that $x = s^2$ for some $s \in S$, and $x \in S$ means that $x = 2n + 1$ for some $n \in \mathbb{Z}$. So our plan (a few minutes later) might look like

$$x \in T,$$

$$x = s^2, \text{ for some } s \in S,$$

smaller space

$$x = 2n + 1, \text{ for some } n \in \mathbb{Z},$$

$$x \in S.$$

We keep filling things in, making sure that each line follows logically from the previous one, until we see how to complete the proof. Here's what we end up with.

Proof.

(Inclusion) Let $x \in T$. Then $x = s^2$ for some $s \in S$. By the definition of S, there exists $n \in \mathbb{Z}$ such that $s = 2n+1$. Hence $x = s^2 = (2n+1)^2 = 4n^2 + 4n + 1 = 2(2n^2 + 2n) + 1$. Now let $m = 2n^2 + 2n$. Then $m \in \mathbb{Z}$ and $x = 2m + 1$. Therefore $x \in S$. Thus $T \subseteq S$, as desired.

(Proper subset) In fact, T is a proper subset of S. To show this we need to exhibit an element that is in S, but not in T. Consider the number -1. Then $-1 = 2(-1) + 1$ and $-1 \in \mathbb{Z}$. Thus, -1 satisfies the defining property for S, so $-1 \in S$. On the other hand, the elements of T are squares of real numbers. Consequently all of them are nonnegative. Hence $-1 \notin T$, and the inclusion is proper. ∎

If you remember the result of Problem 3.1, then you know that you already showed that s^2 is odd if and only if s is odd. If you refer the reader to this result (carefully referencing it, so the reader can find it easily) you can significantly shorten the proof of inclusion. Given two proofs written with equal clarity and insight, most people will prefer the shorter of the two. If the reader remembers the result, reading it again may detract from the proof. So, as long as you tell the reader what you are using and where to find it if he or she needs to, you can (and should) refer to previous results. ○

We now return to the subject of this chapter. Notice that we just told you how to show that a set A is contained in a set B. All you need to do is show that *for all x, if $x \in A$, then $x \in B$*. So we also just told you how to show that A is not contained in B—negate the definition of containment.

Exercise 6.6.

Negate the statement: "For all x, if $x \in A$, then $x \in B$." ○

Two sets are equal if they have precisely the same elements. This can be defined a bit more formally as follows. A set A is **equal** to B, written $A = B$, if $A \subseteq B$ and $B \subseteq A$. To show that two sets are equal is therefore a two step task: First you show that one set is contained in the other ($A \subseteq B$). Then you reverse the order of the sets and show inclusion again ($B \subseteq A$).

Note that when A is a subset of B we use the symbol $\subseteq$, but when x is an element of A we use the symbol $\in$. Choose your symbols carefully and don't mix them up! If x is not in A, then we write $x \notin A$. If A is not a subset of B, then we write $A \nsubseteq B$.

Exercise 6.7.
Write a definition of set equality that reverts back to membership in a set, rather than set containment.   ○

Example 6.8.
Show that $\{x \in \mathbb{R} : x^2 - 1 = 0\} = \{1, -1\}$.

According to the definition of equality above, we have to show two separate things. The first is to show that the set on the left is contained in the set on the right. For this part of the proof, we will begin with an arbitrary element $y \in \{x \in \mathbb{R} : x^2 - 1 = 0\}$ and we will try to show that $y = 1$ or $y = -1$. Then we must show that the set on the right is contained in the set on the left. So for this part, we will begin with $y = 1$ or $y = -1$ and try to show that it is in the set on the left.

Proof.
If $y \in \{x \in \mathbb{R} : x^2 - 1 = 0\}$, then $0 = y^2 - 1 = (y-1)(y+1)$. Hence $y = 1$ or $y = -1$, and $y \in \{1, -1\}$. Therefore, $\{x \in \mathbb{R} : x^2 - 1 = 0\} \subseteq \{1, -1\}$.

Now if $y \in \{1, -1\}$, then $y = 1$ or $y = -1$. In either case we get $y^2 = 1$. Hence $y^2 - 1 = 0$ and $y \in \{x \in \mathbb{R} : x^2 - 1 = 0\}$. Therefore $\{1, -1\} \subseteq \{x \in \mathbb{R} : x^2 - 1 = 0\}$.

By the definition of equality of sets, $\{x \in \mathbb{R} : x^2 - 1 = 0\} = \{1, -1\}$.   ■

Exercise 6.9.
Let $A = \{1, 3, 5\}, B = \{3, 4, 6\}, C = \{5\}$, and $D = \{1, 3\}$. Which sets are subsets of the others? For which sets S do we have $1 \in S$? $1 \notin S$? Which sets are not subsets of each other?   ○

Theorem 6.10.
Let A be a set. Then $\emptyset \subseteq A$.

Proof.

We must show that for every x, if $x \in \emptyset$, then $x \in A$. Since there are no elements in the empty set, the antecedent is always false. Therefore the implication is always true, completing the proof. ∎

We will now present a list of very important definitions, using two sets, A and B, to create other sets. Some examples will be presented (briefly) here, and more can be found in the exercises. In what follows, we assume that all variables x belong to a universe, X.

The **union** of A and B is denoted $A \cup B$ and is defined by $A \cup B = \{x : x \in A \text{ or } x \in B\}$. For example, if A is the set of even integers, and B is the set of odd integers, then $A \cup B = \mathbb{Z}$.

The **intersection** of A and B is $A \cap B = \{x : x \in A \text{ and } x \in B\}$. If A and B are two sets such that $A \cap B = \emptyset$, then we say that A and B are **disjoint**. For example, if A is the set of even integers and B is the set of odd integers, then A and B are disjoint.

The **set difference** of B in A is $A \setminus B = \{x \in A : x \notin B\}$. A comment is in order here. We can never look for objects "not in B" unless we know where to start looking. So we use A to tell us where to look for elements not in B. If A is the universe, we will write B^c for $A \setminus B$. This is referred to as the **complement** of B. For example, let A be the set of integers. If $B = \mathbb{Z}^+$, then $A \setminus B$ is the set of elements of A (integers) that are not in B (that are not positive integers). Thus $A \setminus B = \{x \in \mathbb{Z} : x \leq 0\}$. On the other hand, if $A = \mathbb{N}$, then $A \setminus B = \{0\}$.

It is possible to visualize these sets using a representation called a Venn diagram. These diagrams are often helpful in sorting out the relationship between sets. The universe is usually indicated by a rectangle containing the sets. The idea is illustrated in Figures 6.2 and 6.3.

But be careful—pictures can be deceiving. Use the Venn diagram to get your intuition going, but check everything carefully using the techniques we have developed thus far.

Exercise 6.11.

Use the sets in Exercise 6.9 to answer the following questions: What is $A \setminus B$? $A \setminus C$? Which sets are disjoint? If the universe is $\{1, 2, 3, 4, 5, 6\}$, what is A^c? Find $A \cup B$ and $A \cap B$. ○

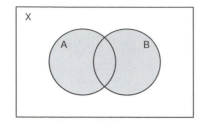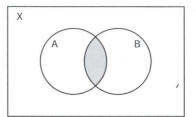

FIGURE 6.2 $A \cup B$ and $A \cap B$

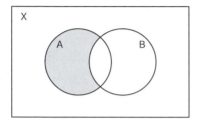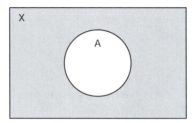

FIGURE 6.3 $A \setminus B$ and A^c

Exercise 6.12.
Write a definition of union for three sets. Write a definition of intersection for three sets. Can you write a definition of set difference for
three sets? Why or why not?   ○

Solutions to Exercises

Solution to Exercise (6.1).
Here is the answer to (b): The universe is the set of all real numbers.
The defining property is "$x \in \mathbb{R} \setminus \mathbb{Q}$."

Solution to Exercise (6.2).
 (a) The adjective "pretty" is subjective and it is unclear whether a
     person from Luxembourg is a member of the set C or not.
 (b) Consider the following question: Is the collection D an element
     of D or not? If it is an element of D, then it must satisfy the

defining property, which says that D is not an element of D; in other words, in this case it would have to be both in the set and not in the set. On the other hand, if D is not an element of the collection D, then it does just what the defining property says. Thus it must be in the set D; in other words, in this case it would have to be both in the set and not in the set. Hence, this property is contradictory.

Solution to Exercise (6.3).
We can write $T = \{x \in \mathbb{Z} : x = (2n + 1)^2 \text{ for some } n \in \mathbb{Z}\}$ and $S = \{2n + 1 : n \in \mathbb{Z}\}$.

Solution to Exercise (6.4).
Let A be the set of nonzero integers.
  (a) $A = \{x \in \mathbb{Z} : x \neq 0\}$.
  (b) Let x and y be elements of A. Then $x \star y = 2xy$. Since x, y, and 2 are all integers, $x \star y \in \mathbb{Z}$. Furthermore, since x and y are elements of A, they are nonzero. Therefore $x \star y = 2xy \neq 0$. Consequently $x \star y \in A$, as desired.
  (c) This is false. Suppose to the contrary that there were such an element y in A. Then $x \star y = x$ for every $x \in A$. Choosing $x = 1$, we see that $1 = 1 \star y = 2(1)(y) = 2y$. The only solution to this equation is $y = 1/2$, which is not an integer and therefore not an element of A. This contradiction shows that no such y can exist.

Solution to Exercise (6.6).
The negation is "There exists an x such that $x \in A$ and $x \notin B$."

Solution to Exercise (6.7).
Two sets A and B are equal if for all x we have $x \in A$ if and only if $x \in B$.

Solution to Exercise (6.9).
The following statements hold:
  $C \subseteq A$ and $D \subseteq A$;
  no other sets are subsets of each other;
  $1 \in A$, $1 \in D$, $1 \notin B$, and $1 \notin C$.

Solution to Exercise (6.11).
The following statements hold:

$A \setminus B = \{1, 5\}$, $A \setminus C = \{1, 3\}$;

sets B and C are disjoint, and sets C and D are disjoint;

if the universe is as given, then $A^c = \{2, 4, 6\}$;

$A \cup B = \{1, 3, 4, 5, 6\}$, and $A \cap B = \{3\}$.

Solution to Exercise (6.12).
Let A, B, and C be sets and let the universe be denoted by X. Then $A \cup B \cup C = \{x \in X : x \in A \text{ or } x \in B \text{ or } x \in C\}$ and $A \cap B \cap C = \{x \in X : x \in A \text{ and } x \in B \text{ and } x \in C\}$. While the union and intersection of three sets makes sense, the set difference of three sets does not. In order to answer this question, we would need to reduce it to a set difference of two sets by including parentheses. For example, you can define the following set differences: $(A \setminus B) \setminus C$ and $A \setminus (B \setminus C)$ (try it!). Work out what these last two sets are when A, B, and C are as in Exercise 6.9.

Spotlight: Paradoxes

You may already have seen paradoxes in mathematics. For example, you may have seen Zeno's paradoxes in your calculus class. Another well-known paradox comes from the following: what is the sum of

$$1 - 1 + 1 - 1 + 1 - \cdots?$$

You might argue that this sum should be $(1 - 1) + (1 - 1) + \cdots = 0$. Or, you might just as well argue that this sum should be $1 + (-1 + 1) + (-1 + 1) + \cdots = 1$. You might even argue (as Luigi Guido Grandi did [23, p. 135]) that since the sums 0 and 1 are equally probable, the answer should be the average of 0 and 1; in other words, $1/2$. This paradox forces us to look closely at exactly what we mean by summing infinitely many numbers.

Betrand Russell pointed out a paradox in set theory. He also presented a popular form of this paradox, called the barber problem. The problem is the following. Suppose there is a town with one barber, and this barber says that he shaves those people, and only those, who do not shave themselves. The question is: Who shaves the barber? (You'll recognize the set theoretic form of this problem in Exercise 6.2.)

Paradoxes serve a very useful purpose. They point out where the foundations of mathematics are shaky (or even faulty!). To learn more about them, and how they have been handled, we recommend reading [22, Chapter 15], [46, Chapter 18], or [48, Chapter 51].

Problems

Problem 6.1.
Recall that $\mathbb{N}$ denotes the set of natural numbers, $\mathbb{Z}$ the set of integers, and $\mathbb{R}$ the set of real numbers.
  (a) Write the phrase "x belongs to $\mathbb{R}$" in symbols.
  (b) Write the phrase "$\mathbb{Z}$ is a proper subset of $\mathbb{R}$" in symbols.
  (c) Write the phrase "If x is an element of $\mathbb{Z}$, then x or $-x$ is an element of $\mathbb{N}$" in symbols.
  (d) Use set notation to describe the set of squares of all multiples of 3.

Problem 6.2.
In this problem our universe is $\mathbb{R}$, the set of real numbers.
  (a) Give an example of subsets A and B of $\mathbb{R}$ that are disjoint.
  (b) Give an example of subsets A and B of $\mathbb{R}$ that are not disjoint and find $A \setminus B$ and $B \setminus A$.
  (c) Give an example of subsets A and B of $\mathbb{R}$ such that $A \subseteq B$.
  (d) Give an example of subsets A, B, and C of $\mathbb{R}$ such that $A \cup (B \cap C) \neq (A \cup B) \cup (A \cup C)$.

Problem 6.3.
The universe in this problem is $\mathbb{R}$. Let A be the closed interval $[0, 2]$ and let B be the closed interval $[-1, 1]$. Find $A \setminus B$, $B \setminus A$, A^c, B^c, $A^c \cap B^c$, $(A \cup B)$, and $(A \cup B)^c$.

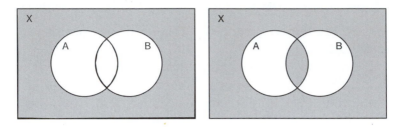

FIGURE 6.4

Problem 6.4.
Find an expression for each of the shaded sets in the Venn diagrams of Figure 6.4

Problem 6.5.
   (a) Consider the set S of nonzero real numbers. Write S in set notation.
   (b) Define a new "multiplication" on this set by $x \heartsuit y = x/y$. If $x, y \in S$, is $x \heartsuit y \in S$? Is there an element $y \in S$ such that $x \heartsuit y = x$ for all $x \in S$?
   (c) Repeat parts (a) and (b), replacing the set S by the set T of negative real numbers.
   (d) Repeat parts (a) and (b), replacing the set S by the set V of nonzero rational numbers.

Problem 6.6.
Define two sets A and B as follows: $A = \{(2n + 1)^3 : n \in \mathbb{Z}\}$ and $B = \{2n + 1 : n \in \mathbb{Z}\}$.
   (a) Prove that $A \subset B$.
   (b) Suppose we redefine A and B, replacing $\mathbb{Z}$ by $\mathbb{R}$; in other words, $A = \{(2n + 1)^3 : n \in \mathbb{R}\}$ and $B = \{2n + 1 : n \in \mathbb{R}\}$. What is the relation between these two sets? State and prove your answer.

Problem 6.7.
Find an expression for each of the shaded sets in the Venn diagrams of Figure 6.5.

Problem 6.8.
Is the following statement true or false: $\{\emptyset\} = \emptyset$? Why?

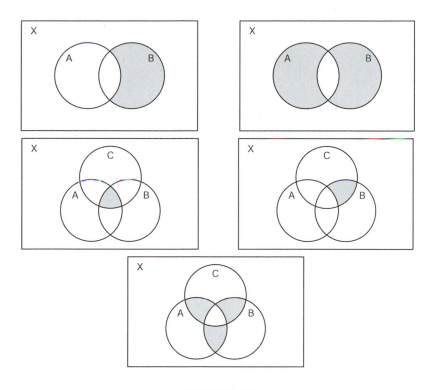

FIGURE 6.5

Problem 6.9.
Let $A = \{x \in \mathbb{Z} : 6 \text{ divides } x\}$, $B = \{x \in \mathbb{Z} : 21 \text{ divides } x\}$ and $C = \{x \in \mathbb{Z} : 42 \text{ divides } x\}$. Prove that $A \cap B = C$.

Problem 6.10.
Let $A = \{(x, y) \in \mathbb{R}^2 : x - y = 0\}$, $B = \{(x, y) \in \mathbb{R}^2 : x + y = 0\}$ and $C = \{(x, y) \in \mathbb{R}^2 : x^2 - y^2 = 0\}$. Prove that $A \cup B = C$.

Problem 6.11.
Let $A = \mathbb{Z}$, $B = \{x \in \mathbb{Z} : x = 2n + 5 \text{ for some } n \in \mathbb{Z}\}$ and $C = \{x \in \mathbb{Z} : x = -2m \text{ for some } m \in \mathbb{Z}\}$. Prove that $A \setminus B = C$.

Problem 6.12.
Let S be the set of nonzero real numbers. Define a new "addition" on this set by $x \sharp y = x + y + 1$. Suppose you add two numbers in

S, do you end up with a number in S? (In other words, if $x, y \in S$, is $x \sharp y \in S$?)

Problem 6.13.
Prove that $A = B$ in each of the following.
  (a) Let A and B be the sets defined by $A = \{x \in \mathbb{R} : \sin(\pi x) = 0\}$ and $B = \mathbb{Z}$.
  (b) Let $x \in \mathbb{R}$. Define the sets A and B by $A = \{(ax + b)/(cx + d) : a, b, c, d \in \mathbb{Z} \text{ and } cx + d \neq 0\}$ and $B = \{(px + q)/(rx + s) : p, q, r, s \in \mathbb{Q} \text{ and } rx + s \neq 0\}$.

Problem 6.14.
Let $A = \{x \in \mathbb{R} : ax^2 + bx + c = 0 \text{ for some integers } a, b, \text{ and } c, \text{ with at least one of } a, b, c \text{ nonzero}\}$ and let $B = \{x \in \mathbb{R} : px^2 + qx + r = 0 \text{ for some rational numbers } p, q, \text{ and } r, \text{ with at least one of } p, q, r \text{ nonzero}\}$.
  (a) Prove that $2 \in A$.
  (b) Prove that $\sqrt{2} \in A$.
  (c) Give an example of a real number y such that $y \notin A$. (You do not need to prove that $y \notin A$.)
  (d) Prove that $A = B$.
  (e) Prove that $\mathbb{Q} \subseteq A$.

   The following problems deal with sets of points in the plane. We remind you of the notation introduced in Chapter 4. The set of all points in the plane is denoted by $\mathbb{R}^2 = \{(x, y) : x, y \in \mathbb{R}\}$. These types of sets will be studied in more generality in Chapter 9.

Problem 6.15.
Define a set A by $A = \{(x, y) \in \mathbb{R}^2 : y \neq 0\}$.
  (a) Give a geometric description of A.
  (b) Suppose we tell you that if you have two elements of this set A, you can "add" them according to the following rule:

$$(x, y) \diamond (z, w) = (xw + zy, wy).$$

  The symbol $+$ here denotes usual addition. Show that the object that results when we add two elements of our set A is again an object in our set A.

(c) Continuing, find an element (a, b) in A such that $(a, b) \diamond (x, y) = (x, y)$ for every (x, y) in A.

(d) This "new" addition probably looks somewhat odd to you, but you have seen it before. What is it?

Problem 6.16.
In each part of this problem, two sets, A and B, are defined. Prove that $A \subseteq B$ in each of the following:
(a) $A = \{x^2 : x \in \mathbb{Z}\}$ and $B = \mathbb{Z}$;
(b) $A = \mathbb{R}$ and $B = \{2x : x \in \mathbb{R}\}$;
(c) $A = \{(x, y) \in \mathbb{R}^2 : y = (5 - 3x)/2\}$ and $B = \{(x, y) \in \mathbb{R}^2 : 2y + 3x = 5\}$.

Problem 6.17.
Prove that one set is a proper subset of the other one in each of the following:
(a) $A = \{(x, y) \in \mathbb{R}^2 : xy > 0\}$ and $B = \{(x, y) \in \mathbb{R}^2 : x^2 + y^2 > 0\}$;
(b) $A = \emptyset$ and $B = \{(x, y) \in \mathbb{R}^2 : x^2 + y^2 \leq 0\}$.

Problem 6.18.
Are the sets

$$\{(x, y) \in \mathbb{R}^2 : x^2 + y^2 \leq 1\} \quad \text{and} \quad \{(x, y) \in \mathbb{R}^2 : |x| + |y| \leq 1\}$$

equal? Justify your answer.

7

CHAPTER

Operations on Sets

By an operation on sets we mean the construction of a new set from the given ones. As we saw in the last chapter, these new sets may be formed using unions, intersections, set differences, or complements of given sets. In this section, we will look at many important properties of operations on sets. We end the chapter with a summarizing list of identities. In the exercises and problems you will be given the opportunity to prove the most important ones and then commit them to memory, so you don't have to re-prove them every time you need them.

The Venn diagrams introduced in the previous chapter can be helpful in deciding what is true and what is false, and they can be part of understanding the problem. All we ask is that you continue to bear in mind that a Venn diagram never constitutes a proof. When you prove these properties you may not always need to start from the definition. Sometimes you can use what you know, and once you have proven everything in Theorem 7.4, you will know a lot.

The first theorem is a good example of a proof in cases. It keeps things tidy. Now remember, if we use the definition to show two sets A and B are equal, then we must show that if $x \in A$, then $x \in B$ *and* if $x \in B$, then $x \in A$.

Theorem 7.1 (The distributive property).
Let A, B, and C be sets. Then $A \cup (B \cap C) = (A \cup B) \cap (A \cup C)$.

   Before reading the proof, let's use Pólya's method.
   "Understanding the problem." Draw two Venn diagrams representing the left and right sides of the equality above. Each diagram will have three sets, appropriately labeled A, B, and C. Shade in the area described by the left side of the equation in one diagram and then shade the right side in the other diagram. They should look the same. While this should convince you that you are on the right track, it is not enough to convince someone else.
   "Devising a plan." We wish to show that two sets are equal. Using the definition of equality of sets, we know that we must show two things. The first thing to show is that $A \cup (B \cap C) \subseteq (A \cup B) \cap (A \cup C)$. So our first line will begin

$$\text{If } x \in A \cup (B \cap C),$$

and our last line (for this part of the proof) will look like

$$\text{Thus } x \in (A \cup B) \cap (A \cup C).$$

Now we just have to figure out how to get from the first line to the last one. Let's fill in some things, making sure that each line follows logically from the previous one. Working down from the top we get

$$x \in A \cup (B \cap C),$$

$$x \in A \text{ or } x \in B \cap C,$$

and working up from the bottom leads to

$$x \in A \cup B \text{ and } x \in A \cup C,$$

$$x \in (A \cup B) \cap (A \cup C).$$

Looking at what we are missing in our proof suggests that we use a proof in cases; one that depends on whether $x \in A$ or $x \in B \cap C$.
   Once we are done with the proof above, we must show that $(A \cup B) \cap (A \cup C) \subseteq A \cup (B \cap C)$. We use the same method to devise our plan for a proof of this set containment: We write down our first line

and look to see where it takes us. Then we'll write down our last line and try to figure out how to get there. That leads to

$$x \in (A \cup B) \cap (A \cup C),$$

$$x \in A \cup B \text{ and } x \in A \cup C,$$

[stuff]

$$x \in A \text{ or } x \in B \cap C,$$

$$x \in A \cup (B \cap C).$$

It looks like if $x \in A$, we have our proof. But what if $x \notin A$? This again suggests a proof in cases; one that depends on whether $x \in A$ or $x \notin A$. If you see what to do now, you can write up the proof. If you still do not see what to do, continue using this method until you see the solution.

Once you see the solution, fill in the missing steps and write the proof up carefully using complete sentences, as we do below.

Proof.
If $x \in A \cup (B \cap C)$, then $x \in A$ or $x \in B \cap C$. Suppose first that $x \in A$. Then $x \in A \cup B$ and $x \in A \cup C$. In this first case, we see that $x \in (A \cup B) \cap (A \cup C)$. Now suppose that $x \in B \cap C$. Then $x \in B$ and $x \in C$. Since $x \in B$, we see that $x \in A \cup B$. Since we also have $x \in C$, we see that $x \in A \cup C$. Therefore, $x \in (A \cup B) \cap (A \cup C)$ in this case as well. In either case $x \in (A \cup B) \cap (A \cup C)$ and we may conclude that $A \cup (B \cap C) \subseteq (A \cup B) \cap (A \cup C)$.

To complete the proof, we must now show that $(A \cup B) \cap (A \cup C) \subseteq A \cup (B \cap C)$. So if $x \in (A \cup B) \cap (A \cup C)$, then $x \in A \cup B$ and $x \in A \cup C$. It is, once again, helpful to break this into two cases, since we know that either $x \in A$ or $x \notin A$. Now if $x \in A$, then $x \in A \cup (B \cap C)$. If $x \notin A$, then the fact that $x \in A \cup B$ implies that x must be in B. Similarly, the fact that $x \in A \cup C$ implies that x must be in C. Therefore, $x \in B \cap C$. Hence $x \in A \cup (B \cap C)$. In either case $x \in A \cup (B \cap C)$ and we may conclude that $(A \cup B) \cap (A \cup C) \subseteq A \cup (B \cap C)$.

Since we proved containment in both directions we may conclude that the two sets are equal. ■

Look at the proof above. It has complete sentences, variables are identified, we know when we are in one case and then the other, and we know when the proof is complete. You should use the form as a model, but remember that each proof will be unique.

We now come to our first proof involving an "if and only if" statement. Remember that an "if and only if" statement requires that you prove *both* the "if" *and* the "only if."

Theorem 7.2.

Let A and B be sets. Then $A \cup B = A$ if and only if $B \subseteq A$.

Proof.

First we'll show that if $A \cup B = A$, then $B \subseteq A$. So assume $A \cup B = A$. If $x \in B$, then $x \in A \cup B$. Using the assumption that $A \cup B = A$ we have $x \in A$. This shows that $B \subseteq A$.

Now we will prove that if $B \subseteq A$, then $A \cup B = A$. So let us assume that $B \subseteq A$. We must show that $A \cup B \subseteq A$ and $A \subseteq A \cup B$. To prove the first containment, we have that if $x \in A \cup B$, then $x \in A$ or $x \in B$. If $x \in A$, then x is where it needs to be and we have nothing more to prove. If $x \in B$, then we use the assumption that $B \subseteq A$ to conclude that $x \in A$. In either case we get $x \in A$ and therefore have $A \cup B \subseteq A$. To prove the second containment, let $x \in A$. Then $x \in A \cup B$ and we conclude that $A \subseteq A \cup B$. Together we have proven that $A \cup B = A$. ∎

The structure of the proof of Theorem 7.2 is more complicated than the proof of the distributive property. First, as we said above, there are two things to prove: the "if" and the "only if." Next, both of these statements have hypotheses and conclusions. In each case, you must be aware of what you are assuming and what you are proving. What's even more important, though, is that you *use* what you are assuming to get to your desired conclusion. If you don't use your assumption, either your original statement was poorly constructed, you proved more than you thought you did, or your proof was in error. In fact, in the proof above, we did not use our assumption that $B \subseteq A$ to prove $A \subseteq A \cup B$. Did we make an error, or did we prove more than we said we did?

Now that you have seen two examples of how to write such a proof, it is time for you to try it by yourself. Try proving one of the two DeMorgan's laws below.

Exercise 7.3.
Let A and B be subsets of the set X. Then

$$X \setminus (A \cup B) = (X \setminus A) \cap (X \setminus B).$$

(a) Devise your plan. (Include a Venn diagram.)
(b) Write up your proof.   ○

We now give the promised list of some of the properties of set operations. We proved three of them above. In the problems you will be asked to work more of the proofs.

Theorem 7.4.
Let X denote a set, and A, B, and C denote subsets of X. Then
1. $\emptyset \subseteq A$ *and* $A \subseteq A$.
2. $(A^c)^c = A$.
3. $A \cup \emptyset = A$.
4. $A \cap \emptyset = \emptyset$.
5. $A \cap A = A$.
6. $A \cup A = A$.
7. $A \cap B = B \cap A$. *(Commutative property)*
8. $A \cup B = B \cup A$. *(Commutative property)*
9. $(A \cup B) \cup C = A \cup (B \cup C)$. *(Associative property)*
10. $(A \cap B) \cap C = A \cap (B \cap C)$. *(Associative property)*
11. $A \cap B \subseteq A$.
12. $A \subseteq A \cup B$.
13. $A \cup (B \cap C) = (A \cup B) \cap (A \cup C)$. *(Distributive property)*
14. $A \cap (B \cup C) = (A \cap B) \cup (A \cap C)$. *(Distributive property)*
15. $X \setminus (A \cup B) = (X \setminus A) \cap (X \setminus B)$. *(DeMorgan's law)*
    (When X is the universe we also write $(A \cup B)^c = A^c \cap B^c$.)
16. $X \setminus (A \cap B) = (X \setminus A) \cup (X \setminus B)$. *(DeMorgan's law)*
    (When X is the universe we also write $(A \cap B)^c = A^c \cup B^c$.)
17. $A \setminus B = A \cap B^c$.

18. $A \subseteq B$ if and only if $(X \setminus B) \subseteq (X \setminus A)$.
    (When X is the universe we also write $A \subseteq B$ if and only if $B^c \subseteq A^c$.)
19. $A \cup B = A$ if and only if $B \subseteq A$.
20. $A \cap B = B$ if and only if $B \subseteq A$.

Many results can be proved using the methods demonstrated thus far in this chapter. Once you have proven these statements, though, it is a good idea to use them in other proofs. Practice using the results in Theorem 7.4 in the next exercise.

Exercise 7.5.
Let A, B, and C be sets. Prove the following using relevant statements from Theorem 7.4: If $C^c \subset B$, then $(A \setminus B) \cup C = C$. $\bigcirc$

Solutions to Exercises

Solution to Exercise (7.3).
First we show that

$$X \setminus (A \cup B) \subseteq (X \setminus A) \cap (X \setminus B).$$

If $x \in X \setminus (A \cup B)$, then $x \notin A \cup B$. Therefore $x \notin A$ and $x \notin B$. Consequently, $x \in X \setminus A$ and $x \in X \setminus B$. Thus $x \in (X \setminus A) \cap (X \setminus B)$. We conclude that $X \setminus (A \cup B) \subseteq (X \setminus A) \cap (X \setminus B)$.
   We now show that

$$(X \setminus A) \cap (X \setminus B) \subseteq X \setminus (A \cup B).$$

If $x \in (X \setminus A) \cap (X \setminus B)$, then $x \in X \setminus A$ and $x \in X \setminus B$. Thus, $x \in X$ and $x \notin A$, and $x \in X$ and $x \notin B$. So, $x \in X$ and $x \notin A$ and $x \notin B$. This implies that $x \in X$ and $x \notin A \cup B$. Therefore, $x \in X \setminus (A \cup B)$, and we see that $(X \setminus A) \cap (X \setminus B) \subseteq X \setminus (A \cup B)$. Thus, the two sets are equal.

Solution to Exercise (7.5).
Since $C^c \subseteq B$, statements 18 and 2 of Theorem 7.4 imply that $B^c \subseteq C$, and thus $B^c \cup C = C$ by statement 19 of the same theorem. The rest of the proof now follows from the following string of equalities

(numbers indicate the relevant statements from Theorem 7.4):

$$
\begin{aligned}
(A \setminus B) \cup C &= (A \cap B^c) \cup C && \text{(by 17)} \\
&= (A \cup C) \cap (B^c \cup C) && \text{(by 8 and 13)} \\
&= (A \cup C) \cap C && \text{(since } B^c \cup C = C \text{ as shown)} \\
&= C && \text{(by 8, 12, and 20).}
\end{aligned}
$$

Problems

In all the problems below, X denotes a set; A, B, and C denote subsets of X.

Problem 7.1.
In this problem we refer to statements of Theorem 7.4.
- (a) Prove statement 2.
- (b) Prove statement 14.
- (c) Prove statement 16.
- (d) Prove statement 18.
- (e) Prove statement 20.

Problem 7.2.
Prove that $A \cap B = \emptyset$ if and only if $B \subseteq (X \setminus A)$.

Problem 7.3.
Prove that $A = B$ if and only if $(X \setminus A) = (X \setminus B)$. Make sure you use statements from Theorem 7.4 rather than going back to the definition.

Problem 7.4.
Prove the following using the results stated in Theorem 7.4:
- (a) $(A \cup B) \cap B = B$;
- (b) $(A \cap B) \cup B = B$.

Problem 7.5.
Prove that $(A \cup B) \setminus (A \cap B) = (A \setminus B) \cup (B \setminus A)$.

Problem 7.6.
Sketch Venn diagrams of the set on the left and the set on the right side of the equation

$$(A \setminus (B \cap C)) \cup (B \setminus C) = (A \cup B) \setminus (B \cap C).$$

Once you have done that, prove that the equality above holds.

Problem 7.7.
Consider the following sets:
  (i) $A \setminus (A \cup B \cup C)$,
  (ii) $A \setminus A \cap B \cap C$,
  (iii) $A \cap B^c \cap C^c$,
  (iv) $A \setminus (B \cup C)$, and
  (v) $(A \setminus B) \cap (A \setminus C)$.
  (a) Which of the sets above are written ambiguously, if any?
  (b) Of the ones that make sense, which of the sets above agree with the shaded set in Figure 7.1?
  (c) Prove that $A \setminus (B \cup C) = (A \setminus B) \cap (A \setminus C)$.

Problem 7.8.
Consider the following sets:
  (i) $(A \cap B) \setminus (A \cap B \cap C)$,
  (ii) $A \cap B \setminus (A \cap B \cap C)$,
  (iii) $A \cap B \cap C^c$,
  (iv) $(A \cap B) \setminus C$, and
  (v) $(A \setminus C) \cap (B \setminus C)$.
  (a) Which of the sets above are written ambiguously, if any?

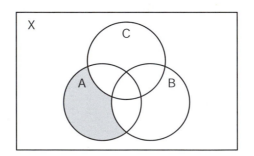

FIGURE 7.1

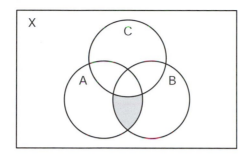

FIGURE 7.2

(b) Of the sets above that make sense, which ones equal the set sketched in Figure 7.2?

(c) Prove that $(A \cap B) \setminus C = (A \setminus C) \cap (B \setminus C)$.

Problem♭ 7.9.

In this problem you will prove that the union of two sets can be rewritten as the union of two disjoint sets.

(a) Prove that the two sets $A \setminus B$ and B are disjoint.

(b) Prove that $A \cup B = (A \setminus B) \cup B$.

Problem 7.10.

Prove or disprove: If $A \cup B = A \cup C$, then $B = C$.

Problem 7.11.

Prove or give a counterexample for the following statement.

Let X be the universe and $A, B \subseteq X$. If $A \cap Y = B \cap Y$ for all $Y \subseteq X$, then $A = B$.

8

CHAPTER

More on Operations on Sets

Most of what we did in the last two chapters was concerned with operations on two sets. In Exercise 6.12 we defined unions and intersections of three sets. In general, we may have two or three sets, as many sets as there are integers, or even more sets than that. We'll need a new definition and special notation. In this chapter, we will introduce the notation that will allow us to keep track of these sets. Unfortunately, a rigorous definition will have to wait until Chapter 13.

Let n be a positive integer and suppose that we have sets $A_1, A_2, \ldots, A_n$. How can we talk about the union of these n sets? the intersection? For example, when we have three sets, if we write $\bigcup_{j=1}^{3} A_j = A_1 \cup A_2 \cup A_3$, we would be referring to the set of x in our universe that lie in at least one of our sets, A_1, A_2, or A_3. Of course, there is nothing special about three sets; that is, for every positive integer, n, we can write

$$\bigcup_{j=1}^{n} A_j = A_1 \cup A_2 \cup \cdots \cup A_n \text{ and } \bigcap_{j=1}^{n} A_j = A_1 \cap A_2 \cap \cdots \cap A_n.$$

The first set would be the set of all x in the universe that lie in at least one of the A_j for $j = 1, 2, \ldots, n$, while the second would be the set of x that lie in all of the sets A_j. If we have a set A_j for each positive integer j and we want to take the union and intersection over all positive integers, then we write

$$\bigcup_{j=1}^{\infty} A_j = A_1 \cup A_2 \cup \cdots \text{ and } \bigcap_{j=1}^{\infty} A_j = A_1 \cap A_2 \cap \cdots.$$

This is probably a good time to look at some examples.

Example 8.1.
We can write the union in different ways. For example, $\bigcup_{j=1}^{10}[0,j] = [0,1] \cup [0,2] \cup \cdots \cup [0,10] = [0,10]$. Similarly, $\bigcap_{j=1}^{10}[0,j] = [0,1] \cap [0,2] \cap \cdots \cap [0,10] = [0,1]$. ○

In Example 8.1, we had unions and intersections of finitely many sets (ten, to be precise). We now take a look at what can happen when we take unions and intersections of even more sets.

Example 8.2.
(a) For each $q \in \mathbb{Z}^+$ define the set $A_q = \{p/q : p \in \mathbb{Z}\}$. These sets can be used to define the union $\bigcup_{q \in \mathbb{Z}^+} A_q$.
(b) This time we define, for each $i \in \mathbb{N}$, the set $B_i = \{p/3^i, p \in \mathbb{Z}\}$. These sets may be used to define the intersection $\bigcap_{i \in \mathbb{N}} B_i$. ○

Exercise 8.3.
Write the sets $\bigcup_{j=1}^{\infty}[j,j+1]$ and $\bigcap_{j=1}^{\infty}[j,j+1]$ in their simplest form, by listing the first few sets in the union or intersection until the pattern is established, and then stating your guess. (You don't have to prove that your guess is correct.) ○

Sometimes we do not know how many sets we have. While this may seem odd, it happens all the time. So suppose we have a set I, and suppose further that for each $\alpha \in I$ there is a set A_α corresponding to it. The set I is called an **index set**, each $\alpha \in I$ is called an index, and the set $\{A_\alpha : \alpha \in I\}$ is called an **indexed family** of sets. We may also write $\{A_\alpha\}_{\alpha \in I}$, and we will often refer to $\{A_\alpha\}_{\alpha \in I}$ as a collection of sets or a family of sets.

We are now ready for the general definition of a union and intersection of sets. Let X denote our universe and let $\{A_\alpha : \alpha \in I\}$ be a family of sets with $A_\alpha \subseteq X$ for all α in an index set I. Then the **union of the family** $\{A_\alpha : \alpha \in I\}$ is defined by

$$\bigcup_{\alpha \in I} A_\alpha = \{x \in X : x \in A_\alpha \text{ for some } \alpha \in I\},$$

and for $I \neq \emptyset$, the **intersection of the family** $\{A_\alpha : \alpha \in I\}$ is defined by

$$\bigcap_{\alpha \in I} A_\alpha = \{x \in X : x \in A_\alpha \text{ for all } \alpha \in I\}.$$

Exercise 8.4.
Find the simplest way to describe the following sets (you may find sketches helpful):
  (a) $\bigcup_{x \in \mathbb{R}^+}(0, x)$;
  (b) $\bigcup_{n \in \mathbb{N}}[0, n]$;
  (c) $\bigcap_{n \in \mathbb{N}}[0, n]$.

Note that the index notation and the general definition of union and intersection given here include the cases in Chapter 6 and the ones we mentioned in the beginning of this chapter. For instance, if $I = \{1, 2\}$, then $\bigcap_{i \in I} A_i = A_1 \cap A_2$.
  Some more practice with this notation will probably be very helpful at this point.

Exercise 8.5.
  (a) Write $\bigcup_{j=0}^{\infty}[0, j]$ using an appropriate index set.
  (b) Write $\bigcup_{j=1}^{\infty}(0, j)$ using an appropriate index set.

Some sets are more easily described with index notation, others without such notation. Let's go back and look at the sets in Example 8.2.

Example 8.6.
Consider the family of sets $\{A_q\}_{q \in \mathbb{Z}^+}$ defined in Example 8.2 (a). Then $\bigcup_{q \in \mathbb{Z}^+} A_q = \mathbb{Q}$.

Proof.

If $x \in \bigcup_{q \in \mathbb{Z}^+} A_q$, then $x \in A_q$ for some $q \in \mathbb{Z}^+$. Therefore, there exist $q \in \mathbb{Z}^+$ and $p \in \mathbb{Z}$ such that $x = p/q$. Consequently $x \in \mathbb{Q}$, and we have shown that $\bigcup_{q \in \mathbb{Z}^+} A_q \subseteq \mathbb{Q}$.

Conversely, if $x \in \mathbb{Q}$, then $x = p/q$ for some $p, q \in \mathbb{Z}$ with $q \neq 0$. Now (for reasons that you will explain) we may choose q so that $q > 0$. For this q we have $q \in \mathbb{Z}^+$ and therefore $x \in A_q$. Hence $x \in \bigcup_{q \in \mathbb{Z}^+} A_q$. So, $\mathbb{Q} \subseteq \bigcup_{q \in \mathbb{Z}^+} A_q$, and therefore $\bigcup_{q \in \mathbb{Z}^+} A_q = \mathbb{Q}$. ∎

Example 8.7.

Consider the family of sets $\{B_i\}_{i \in \mathbb{N}}$ defined in Example 8.2 (b). We claim that $\bigcap_{i \in \mathbb{N}} B_i = \mathbb{Z}$.

Proof.

If $x \in \bigcap_{i \in \mathbb{N}} B_i$, then $x \in B_i$ for all $i \in \mathbb{N}$. In particular, $x \in B_0$. Hence $x = p/3^0 = p$ for some $p \in \mathbb{Z}$. So $x \in \mathbb{Z}$, and therefore $\bigcap_{i \in \mathbb{N}} B_i \subseteq \mathbb{Z}$.

Now let $x \in \mathbb{Z}$. For each $i \in \mathbb{N}$, we may write $x = (3^i x)/3^i$. Of course, $3^i x \in \mathbb{Z}$, since $x \in \mathbb{Z}$. Hence $x \in B_i$ for all $i \in \mathbb{N}$. This shows that $x \in \bigcap_{i \in \mathbb{N}} B_i$ and therefore $\mathbb{Z} \subseteq \bigcap_{i \in \mathbb{N}} B_i$.

Combining the two arguments we obtain the desired equality, $\bigcap_{i \in \mathbb{N}} B_i = \mathbb{Z}$. ∎

Exercise 8.8.

What's the difference between "an infinite union of sets" and "a union of infinite sets"? Give an example of each, showing how these two phrases differ. (While we haven't given a rigorous definition of infinite here, your intuition should suffice to solve this problem.)○

You already know that one of DeMorgan's laws for two sets can be stated as

$$X \setminus (A \cup B) = (X \setminus A) \cap (X \setminus B).$$

This can be rephrased in words as "the complement of a union is the intersection of the complements." DeMorgan's laws do not depend on the number of sets that we have, and that is the point of the next exercise.

Exercise 8.9.

Show that the general DeMorgan's laws hold: For every universe X, index set I, and indexed family of sets $\{A_\alpha : \alpha \in I\}$, we have

(i) $X \setminus \bigcup_{\alpha \in I} A_\alpha = \bigcap_{\alpha \in I}(X \setminus A_\alpha)$ and

(ii) $X \setminus \bigcap_{\alpha \in I} A_\alpha = \bigcup_{\alpha \in I}(X \setminus A_\alpha)$.

$\circ$

Exercise 8.10.

Suppose $A_\alpha \subseteq B$ for each $\alpha \in I$. Show that $\bigcup_{\alpha \in I} A_\alpha \subseteq B$.

$\circ$

Solutions to Exercises

Solution to Exercise (8.3).

You can see that $\bigcup_{j=1}^{\infty}[j, j+1] = [1, 2] \cup [2, 3] \cup [3, 4] \cup \cdots = [1, \infty)$ and $\bigcap_{j=1}^{\infty}[j, j+1] = [1, 2] \cap [2, 3] \cap [3, 4] \cap \cdots = \emptyset$.

Solution to Exercise (8.4).

You can check the following:

(a) $\bigcup_{x \in \mathbb{R}^+}(0, x) = (0, \infty)$;

(b) $\bigcup_{n \in \mathbb{N}}[0, n] = [0, \infty)$;

(c) $\bigcap_{n \in \mathbb{N}}[0, n] = \{0\}$.

Solution to Exercise (8.5).

You can check the following:

(a) $\bigcup_{j=0}^{\infty}[0, j] = \bigcup_{j \in \mathbb{N}}[0, j]$;

(b) $\bigcup_{j=1}^{\infty}(0, j) = \bigcup_{j \in \mathbb{Z}^+}(0, j)$.

Solution to Exercise (8.8).

An infinite union of sets would mean that we take the union over infinitely many sets (but the sets themselves may be finite); in other words, the index set is infinite. For example, $\bigcup_{n \in \mathbb{N}}\{n\}$ would be an infinite union of (finite) sets. On the other hand, a union of infinite sets would mean that the sets themselves must be infinite (while the index set may be finite). For example, the union of the even integers, $2\mathbb{Z}$, with the odd integers, $2\mathbb{Z} + 1$, would be a union of two infinite sets.

Solution to Exercise (8.9).

We will show part (*i*) and will leave part (*ii*) for you to do. So we need to show that

$$X \setminus \left(\bigcup_{\alpha \in I} A_\alpha \right) = \bigcap_{\alpha \in I} (X \setminus A_\alpha).$$

If $x \in X \setminus (\bigcup_{\alpha \in I} A_\alpha)$, then $x \in X$ and $x \notin \bigcup_{\alpha \in I} A_\alpha$. By the definition of union this means that $x \in X$ and $x \notin A_\alpha$ for every $\alpha \in I$. Hence, $x \in X \setminus A_\alpha$ for all $\alpha \in I$, and therefore $x \in \bigcap_{\alpha \in I} (X \setminus A_\alpha)$. Thus, $X \setminus (\bigcup_{\alpha \in I} A_\alpha) \subseteq \bigcap_{\alpha \in I} (X \setminus A_\alpha)$.

Now if $x \in \bigcap_{\alpha \in I} (X \setminus A_\alpha)$, then $x \in X \setminus A_\alpha$ for all $\alpha \in I$. This implies that $x \in X$ and $x \notin A_\alpha$ for every $\alpha \in I$. Hence $x \in X$ and $x \notin \bigcup_{\alpha \in I} A_\alpha$. It follows that $x \in X \setminus (\bigcup_{\alpha \in I} A_\alpha)$ and thus $\bigcap_{\alpha \in I} (X \setminus A_\alpha) \subseteq X \setminus (\bigcup_{\alpha \in I} A_\alpha)$.

The two subset relations give the desired equality between the sets.

Solution to Exercise (8.10).

If $x \in \bigcup_{\alpha \in I} A_\alpha$, then there exists α_0 such that $x \in A_{\alpha_0}$. Since we suppose that $A_{\alpha_0} \subseteq B$, we know that $x \in B$. Thus $\bigcup_{\alpha \in I} A_\alpha \subseteq B$.

Problems

Problem 8.1.

Consider the intervals of real numbers given by $A_n = [0, 1/n)$, $B_n = [0, 1/n]$, and $C_n = (0, 1/n)$.

(a) Find $\bigcup_{n=1}^{\infty} A_n$, $\bigcup_{n=1}^{\infty} B_n$, and $\bigcup_{n=1}^{\infty} C_n$.

(b) Find $\bigcap_{n=1}^{\infty} A_n$, $\bigcap_{n=1}^{\infty} B_n$, and $\bigcap_{n=1}^{\infty} C_n$.

(c) Does $\bigcup_{n \in \mathbb{N}} A_n$ make sense? Why or why not?

Problem 8.2.

If $A_x = [-x, x]$, find $\bigcup_{x \in \mathbb{R}^+} A_x$ and $\bigcap_{x \in \mathbb{R}^+} A_x$.

Problem 8.3.

Find simpler notation for the two sets

$$A = \bigcup_{j=0}^{\infty} [j, j+1] \qquad \text{and} \qquad B = \bigcap_{j \in \mathbb{Z}} (\mathbb{R} \setminus (j, j+1)).$$

Problem 8.4.

Prove or give a counterexample: Let $\{A_n : n \in \mathbb{Z}^+\}$ and $\{B_n : n \in \mathbb{Z}^+\}$ be two indexed families of sets. If $A_n \subset B_n$ for all $n \in \mathbb{Z}^+$, then

$$\bigcap_{n=1}^{\infty} A_n \subset \bigcap_{n=1}^{\infty} B_n.$$

(Recall that $A \subset B$ means strict inclusion; that is, $A \subseteq B$ and $A \neq B$.)

Problem 8.5.

Let $\{A_r : r \in \mathbb{R}\}$ and $\{B_r : r \in \mathbb{R}\}$ be two indexed families of sets. Prove that

$$\left(\bigcap_{r \in \mathbb{R}} A_r\right) \cup \left(\bigcap_{r \in \mathbb{R}} B_r\right) \subseteq \bigcap_{r \in \mathbb{R}} (A_r \cup B_r).$$

Provide an example showing that this inclusion can be proper.

Problem♭ 8.6.

Let $\{A_\alpha : \alpha \in I\}$ be an indexed family of sets, and let B be a set.
   (a) Prove the distributive property:

$$\left(\bigcup_{\alpha \in I} A_\alpha\right) \cap B = \bigcup_{\alpha \in I} (A_\alpha \cap B).$$

   (b) State and prove a distributive property for $\left(\bigcap_{\alpha \in I} A_\alpha\right) \cup B$.

Problem♭ 8.7.

Suppose that $\{A_\alpha : \alpha \in I\}$ is an indexed family of subsets of a set X, and that B is a subset of X.
   (a) If $A_\alpha = \emptyset$ for some $\alpha \in I$, prove that $\bigcap_{\alpha \in I} A_\alpha = \emptyset$.
   (b) If $A_\alpha = X$ for some $\alpha \in I$, prove that $\bigcup_{\alpha \in I} A_\alpha = X$.
   (c) If $B \subseteq A_\alpha$ for every $\alpha \in I$, prove that $B \subseteq \bigcap_{\alpha \in I} A_\alpha$.

Problem 8.8.

Define

$$A = \mathbb{R} \setminus \bigcap_{n \in \mathbb{Z}^+} (\mathbb{R} \setminus \{-n, -n+1, \dots, 0, \dots, n-1, n\}).$$

The set A should be familiar to you. Guess what it is and then prove that your guess is correct.

Problem 8.9.

Guess a simpler way to express the set A defined as

$$A = \mathbb{Q} \setminus \bigcap_{n \in \mathbb{Z}} (\mathbb{R} \setminus \{2n\}),$$

and then prove that your guess is correct.

Problem 8.10.

Suppose that X is a set with more than one element. What is $\bigcup_{x \in X} \{x\}$? What is $\bigcap_{x \in X} \{x\}$?

Problem$^\flat$ 8.11.

A collection of sets $\{A_\alpha : \alpha \in I\}$ is said to be a **pairwise disjoint collection** if the following is satisfied: For all $\alpha, \beta \in I$, if $A_\alpha \cap A_\beta \neq \emptyset$, then $A_\alpha = A_\beta$. Suppose that each set A_α is nonempty.

(a) Give an example of pairwise disjoint sets $A_1, A_2, A_3, \ldots$.
(b) What is the contrapositive of "if $A_\alpha \cap A_\beta \neq \emptyset$, then $A_\alpha = A_\beta$"?
(c) What is the converse of "if $A_\alpha \cap A_\beta \neq \emptyset$, then $A_\alpha = A_\beta$"?
(d) If $\{A_\alpha : \alpha \in I\}$ is a pairwise disjoint collection, does the assertion you found in (b) hold for all α and β in I?
(e) If the assertion that you found in (b) holds for all α and β in I, is $\{A_\alpha : \alpha \in I\}$ a pairwise disjoint collection?
(f) If $\{A_\alpha : \alpha \in I\}$ is a pairwise disjoint collection of sets, does it follow that $\bigcap_{\alpha \in I} A_\alpha = \emptyset$?
(g) If $\bigcap_{\alpha \in I} A_\alpha = \emptyset$, is $\{A_\alpha : \alpha \in I\}$ necessarily a pairwise disjoint collection of sets?

Problem 8.12.

Find an example of sets $\{A_j : j \in \mathbb{Z}^+\}$ such that $A_{j+1} \subset A_j$ for each $j \in \mathbb{Z}^+$, and $\bigcap_{j=1}^{\infty} A_j \neq \emptyset$.

9
CHAPTER

The Power Set and the Cartesian Product

Now that we know about sets, we can construct some new ones from old ones in even more ways than we did before. In this section we look closely at two special sets: the first is called the power set, and the second is called the Cartesian product of two sets.

Let S be a set. Then the **power set** of S is the set of all subsets of S. We shall denote the power set by $\mathcal{P}(S)$. Before we begin, note that the power set is again a set and its elements are also sets. The power set is never empty. Why?

Example 9.1.
Consider the set $S = \{0, 1\}$. Then the power set of S is $\mathcal{P}(S) = \{\emptyset, \{0\}, \{1\}, \{0, 1\}\}$. ○

As you probably noticed, the notation is tricky here.

Exercise 9.2.
Let $A = \{1, 2, 3\}$, $B = \{2, 5\}$, $C = \{0, 1\}$.
  (a) Find $\mathcal{P}(B)$ and $\mathcal{P}(C)$. Do these two sets have elements in common?
  (b) Find $\mathcal{P}(A)$, $\mathcal{P}(B)$, $\mathcal{P}(A \cap B)$, and $\mathcal{P}(A \cup B)$.

(c) Compute $\mathcal{P}(A) \cup \mathcal{P}(B)$ and $\mathcal{P}(A) \cap \mathcal{P}(B)$. ○

Remember to use element notation when you are thinking of the set as an element and subset notation when you are showing containment of sets.

In the next exercise, we will ask you to prove that two sets are equal. We've done this many times in the previous chapters, and so you know one way to begin: use an element-chasing argument. Ask yourself if your set plays the role of a set or the role of an element, and use the corresponding notation.

Exercise 9.3.
Let A and B be sets. Prove that $\mathcal{P}(A \cap B) = \mathcal{P}(A) \cap \mathcal{P}(B)$. ○

When we talk about a set, it is understood that if we discuss the set $\{1, 3\}$ we are discussing the set $\{3, 1\}$ as well. A set is just a collection of objects and there is no notion of order associated with what we have defined so far. When there is an order, such as when we plot points and need to know which is the x coordinate and which is the y coordinate, we use the notion of an ordered pair. The next set we will consider is called the Cartesian product of two sets X and Y, and it is constructed using ordered pairs.

Here is our informal definition: An **ordered pair** (x, y) is a pair of objects in which there is a first object x and a second object y. The very important property of ordered pairs is that $(x, y) = (z, w)$ if and only if $x = z$ and $y = w$.

We may now define the **Cartesian product** of X and Y, denoted $X \times Y$, to be the set of all ordered pairs in which the first element comes from X and the second from Y; that is,

$$X \times Y = \{(x, y) : x \in X, y \in Y\}.$$

For example, if $X = [0, 1]$ and $Y = [0, 2]$, then

$$X \times Y = \{(x, y) : 0 \le x \le 1, \ 0 \le y \le 2\}.$$

This is the rectangle in $\mathbb{R}^2$ with base along the interval $[0, 1]$ and height along the interval $[0, 2]$ sketched in Figure 9.1.

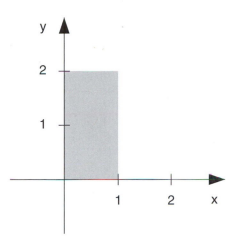

FIGURE 9.1 $[0, 1] \times [0, 2]$

Perhaps you are wondering why we said "informal definition" when we presented our definition of ordered pair. Since this is probably the definition you were expecting, it most likely looks formal. It turns out that there is a rigorous definition of ordered pair; one that can be presented without referring to the "first" and "second" coordinates. The reason we do not present it here is that, in our opinion, a rigorous definition is mostly confusing rather than helpful at this point. If you have a strong desire to know more about this, you can work Problem 9.16 in this chapter.

It's time for a few more examples of Cartesian products.

Exercise 9.4.
(a) Write out all the elements in $\{0, 1\} \times \{2, 3\}$ and $\{2, 3\} \times \{0, 1\}$.
(b) Sketch the Cartesian products $[0, 1] \times [2, 3]$ and $[2, 3] \times [0, 1]$ as sets of points in the plane.
(c) Recall that we defined $\mathbb{R}^2 = \{(x, y) : x \in \mathbb{R}, y \in \mathbb{R}\}$. Write $\mathbb{R}^2$ using the Cartesian product notation.
(d) Having done that, can you describe $\mathbb{R}^3$ as a Cartesian product of two sets? (You might have more than one description that seems reasonable to you.)
(e) We denote the set of even integers by $2\mathbb{Z} = \{2n : n \in \mathbb{Z}\}$. Make sketches that describe the sets $\mathbb{Z} \times \mathbb{Z}$, $\mathbb{Z} \times 2\mathbb{Z}$, and $2\mathbb{Z} \times \mathbb{Z}$. ○

When we prove that two sets defined using Cartesian products are equal, we can still use the method of "element-chasing." However, if we don't use the special form of the element, namely that it looks like an ordered pair, we are likely to get stuck. Notice how we use the special form of the ordered pair in our proofs below.

Theorem 9.5.
Let A be a set. Then $A \times \emptyset = \emptyset$.

Proof.
Suppose, to the contrary, that $A \times \emptyset \neq \emptyset$. Then there exists an element $(x, y) \in A \times \emptyset$. Therefore, by our definition of Cartesian product, $x \in A$ and $y \in \emptyset$. But this contradicts the fact that $\emptyset$ is the empty set. Thus $A \times \emptyset = \emptyset$. ∎

Theorem 9.6.
Let $A, B, C,$ and D be sets. Then

$$(A \times B) \cup (C \times D) \subseteq (A \cup C) \times (B \cup D).$$

Proof.
If $z \in (A \times B) \cup (C \times D)$, then $z = (x, y)$ where $(x, y) \in A \times B$ or $(x, y) \in C \times D$. Suppose first that $(x, y) \in A \times B$. Then $x \in A$ and $y \in B$. In this case $x \in A \cup C$ and $y \in B \cup D$, so by definition $(x, y) \in (A \cup C) \times (B \cup D)$. Now suppose that $(x, y) \in C \times D$. Then $x \in C$ and $y \in D$. Therefore $x \in A \cup C$ and $y \in B \cup D$. So $(x, y) \in (A \cup C) \times (B \cup D)$. Hence $(A \times B) \cup (C \times D) \subseteq (A \cup C) \times (B \cup D)$, as desired. ∎

Again, notice how quickly we changed from z to (x, y) in the proof. That's because we can't do anything if we don't realize that z is really an ordered pair.

Now consider the following nontheorem.

Nontheorem 9.7.
Let $A, B, C,$ and D be sets. Then $(A \cup C) \times (B \cup D) \subseteq (A \times B) \cup (C \times D)$.

Not a proof.
If $(x, y) \in (A \cup C) \times (B \cup D)$, then $x \in A \cup C$ and $y \in B \cup D$. Thus $x \in A$ or $x \in C$, and $y \in B$ or $y \in D$. Hence $x \in A$ and $y \in B$ or $x \in C$ and $y \in D$.

So $(x, y) \in A \times B$ or $(x, y) \in C \times D$. Thus $(x, y) \in (A \times B) \cup (C \times D)$.

$\boxed{?}$

Exercise 9.8.
Find the error in the nonproof above and show that Nontheorem 9.7 really is not a theorem because the statement is false. (Find sets for which the statement does not hold.) ○

In these problems and all that follow, you will begin with an element in your set. It will be helpful to you to think about the form of your element. Is it a set? an ordered pair? If you rush through these proofs, as we did in Nontheorem 9.7, you will prove things that are false. This is generally frowned upon in mathematics. Go slowly, be careful, and check each step.

We will now define relations. We will soon see that there is a connection between functions (something you probably feel familiar with) and relations (something you may not feel terribly familiar with). We begin with a definition.

Suppose that X and Y are two sets. A **relation from** X **to** Y is a subset of $X \times Y$. A relation from X to X is called a **relation on** X.

Exercise 9.9.
For the following, decide whether or not they are relations from a set X to a set Y. If they are, say what X is and what Y is. Then describe each set either pictorially (as a set of points in the plane) or in words:
  (a) $\{(x, y) \in \mathbb{R}^2 : x \leq y\}$;
  (b) $\{x/y : x, y \in \mathbb{Z} \text{ and } y \neq 0\}$;
  (c) $\{(x, y) \in \mathbb{R}^2 : x, y \in \mathbb{Z} \text{ and } x + y = 0\}$. ○

We will learn more about relations in Chapter 10.

Solutions to Exercises

Solution to Exercise (9.2).
  (a) $\mathcal{P}(B) = \{\emptyset, \{2\}, \{5\}, \{2, 5\}\}$, $\mathcal{P}(C) = \{\emptyset, \{0\}, \{1\}, \{0, 1\}\}$, and the empty set is an element of both sets.

(b) $\mathcal{P}(A) = \{\emptyset, \{1\}, \{2\}, \{3\}, \{1, 2\}, \{1, 3\}, \{2, 3\}, \{1, 2, 3\}\}$, $\mathcal{P}(A \cap B) = \{\emptyset, \{2\}\}$. We leave $\mathcal{P}(A \cup B)$ to you.

(c) $\mathcal{P}(A) \cap \mathcal{P}(B) = \{\emptyset, \{2\}\}$. We leave $\mathcal{P}(A) \cup \mathcal{P}(B)$ to you.

Solution to Exercise (9.3).
If $x \in \mathcal{P}(A \cap B)$, then $x \subseteq A \cap B$. This implies that $x \subseteq A$ and $x \subseteq B$. Thus $x \in \mathcal{P}(A)$ and $x \in \mathcal{P}(B)$, so $x \in \mathcal{P}(A) \cap \mathcal{P}(B)$. Since x was arbitrary, $\mathcal{P}(A \cap B) \subseteq \mathcal{P}(A) \cap \mathcal{P}(B)$. Each of these steps is reversible, so the containment $\mathcal{P}(A) \cap \mathcal{P}(B) \subseteq \mathcal{P}(A \cap B)$ follows as well.

Solution to Exercise (9.4).
(a) The two sets are

$$\{0, 1\} \times \{2, 3\} = \{(0, 2), (0, 3), (1, 2), (1, 3)\}$$

and

$$\{2, 3\} \times \{0, 1\} = \{(2, 0), (3, 0), (2, 1), (3, 1)\}.$$

(b) The two sets are sketched in Figure 9.2 below.
(c) $\mathbb{R}^2 = \mathbb{R} \times \mathbb{R}$.
(d) One answer might be $\mathbb{R}^2 \times \mathbb{R}$.
(e) These are sketched in Figure 9.3.

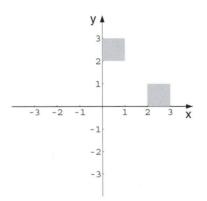

FIGURE 9.2 $[0, 1] \times [2, 3]$ and $[2, 3] \times [0, 1]$

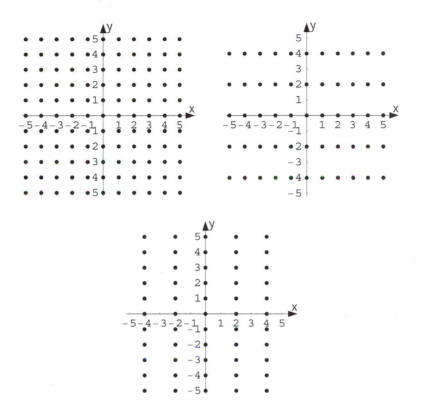

FIGURE 9.3 $\mathbb{Z} \times \mathbb{Z}$, $\mathbb{Z} \times 2\mathbb{Z}$ and $2\mathbb{Z} \times \mathbb{Z}$

Solution to Exercise (9.8).
Our nonproof claims that "$x \in A$ or $x \in C$, and $y \in B$ or $y \in D$. Hence $x \in A$ and $y \in B$ or $x \in C$ and $y \in D$." This conclusion is not justified: it could also be that $x \in A$ and $y \in D$, or $x \in C$ and $y \in B$.

Consider the following example: Let $A = D = \emptyset$ and $B = C = \mathbb{R}$. Then $(A \cup C) \times (B \cup D) = \mathbb{R} \times \mathbb{R}$, while $(A \times B) \cup (C \times D) = \emptyset$.

Solution to Exercise (9.9).
(a) This is a relation from $X = \mathbb{R}$ to $Y = \mathbb{R}$, consisting of the set of points in $\mathbb{R}^2$ lying below the line $x = y$.
(b) This is not a subset of $X \times Y$ for any choice of X and Y.
(c) This is a relation from $X = \mathbb{Z}$ to $Y = \mathbb{Z}$, and consists of the points for which x is an integer and $y = -x$; that is, this is the set $\{(x, -x) : x \in \mathbb{Z}\}$.

Problems

Problem 9.1.
Let $S = \{a, b, c\}$. Find $\mathcal{P}(S)$.

Problem 9.2.
(a) Show that $\mathcal{P}(A) \cup \mathcal{P}(B) \subseteq \mathcal{P}(A \cup B)$.
(b) Show that $\mathcal{P}(A) \cup \mathcal{P}(B) \neq \mathcal{P}(A \cup B)$ by exhibiting two concrete sets, A and B, for which the aforementioned inequality holds.

Problem 9.3.
Let $2\mathbb{Z}$ denote the even integers and $2\mathbb{Z} + 1$ denote the odd integers. What is $\mathcal{P}(2\mathbb{Z}) \cap \mathcal{P}(2\mathbb{Z} + 1)$?

Problem 9.4.
Show that $A \subseteq B$ if and only if $\mathcal{P}(A) \subseteq \mathcal{P}(B)$.

Problem 9.5.
For every set I and for every family of sets $\{A_\alpha : \alpha \in I\}$, prove that

$$\bigcup_{\alpha \in I} \mathcal{P}(A_\alpha) \subseteq \mathcal{P}\left(\bigcup_{\alpha \in I} A_\alpha\right).$$

Problem 9.6.
Let $\{A_\alpha : \alpha \in I\}$ be a nonempty family of sets. Prove that

$$\mathcal{P}\left(\bigcap_{\alpha \in I} A_\alpha\right) = \bigcap_{\alpha \in I} \mathcal{P}(A_\alpha).$$

Problem 9.7.
How many elements are there in the power set of $\{1, 2, 3, 4\}$? How many elements are in the power set of $\{1, 2, 3, 4, 5\}$? State a general result. You'll be able to prove it later.

Problem 9.8.
Describe the following relations pictorially (as a set of points in the plane) or in words:
(a) $\{(x, y) \in \mathbb{N} \times \mathbb{Z} : x \geq y\}$;
(b) $\{(x, y) \in \mathbb{R}^2 : x = y\}$;

(c) $\{(x, y) \in \mathbb{Z} \times \mathbb{Z} : x + y \in 2\mathbb{Z}\}$;

(d) $\{0, 1\} \times \mathbb{N}$;

(e) $\{(x, x^2) : x \in \mathbb{R}\}$;

(f) $\{(\sqrt{x}, x) : x \in \mathbb{Z}^+\}$.

Problem 9.9.

Describe the following Cartesian products:

(a) $\emptyset \times \mathbb{N}$;

(b) $\mathbb{Z} \times \emptyset$;

(c) $\mathbb{R} \times \mathbb{R}$;

(d) $\mathbb{R} \times \mathbb{Z}$.

Problem 9.10.

Show that $\mathbb{N} \times \mathbb{N} \subseteq \mathbb{Z} \times \mathbb{Z}$.

Problem 9.11.

(a) In the plane, sketch the set $[0, 1] \times ([1, 3] \cup [2, 4])$.

(b) Sketch $([0, 1] \cup [1, 4]) \times ([0, 1] \cup [2, 4])$.

Problem 9.12.

(a) Prove the following:

Let A, B, C, and D be nonempty sets. Then $A \times B = C \times D$ if and only if $A = C$ and $B = D$.

(b) Where did your proof use the fact that the sets were nonempty?

Problem 9.13.

Suppose A, B, C, and D are four sets. If $A \times B \subseteq C \times D$, must $A \subseteq C$ and $B \subseteq D$? Why or why not?

Problem 9.14.

Let A, B, and C be sets. If the statements below are true prove them. If they are false, give a counterexample:

(a) $A \times (B \cup C) = (A \times B) \cup (A \times C)$;

(b) $A \times (B \cap C) = (A \times B) \cap (A \times C)$.

Problem 9.15.

Let $A = \{1, \{1\}, \{1, \{1\}\}\}$.

(a) Find $A \times A$.

(b) Find $A \cap \mathcal{P}(A)$.

Problem 9.16.
This problem introduces rigorous definitions of an ordered pair and Cartesian product. Let A be a set and $a, b \in A$. We define the ordered pair of a and b with first coordinate a and second coordinate b as

$$(a, b) = \{\{a\}, \{a, b\}\}.$$

Using this definition prove the following.
  (a) If $(a, b) = (x, y)$, then $a = x$ and $b = y$.
  (b) If $a \in A$ and $b \in B$, then $(a, b) \in \mathcal{P}(\mathcal{P}(A \cup B))$.
Now we are able to define the Cartesian product of the two sets A and B as the set

$A \times B = \{x \in \mathcal{P}(\mathcal{P}(A \cup B)) : x = (a, b) \text{ for some } a \in A \text{ and some } b \in B\}$.

  (c) Use the above definitions to prove that if $A \subseteq C$ and $B \subseteq D$, then $A \times B \subseteq C \times D$.

  This is a pretty complicated definition. It is also not our idea, but rather an idea that was born from axioms. P. Halmos' book, [31], is an excellent reference for this subject.

Tips on Writing Mathematics

> Sorry this letter is so long; I didn't have time to make it shorter.—Mark Twain

  After this point in the course the work will change. You'll find that you are writing more in words than in symbols. How you write is as important as what you write. Here are some things to think about as you write your proofs.
  • In mathematics, it is always important that the reader know what the variables stand for. This was true in algebra in high school, geometry, and calculus, and it is true here too. If you use symbols—any symbols—make sure the meaning is clear to the reader *before* you use them.
  • Think about your notation, and choose notation that is easy on the reader.

- A variable should only be assigned one meaning in your proof. For example, if you used C to denote the complex numbers, don't use C again to denote a set.
- Try for a good blend of symbols and words. Don't juxtapose unrelated symbols if you don't have to. For example, consider the sentence "So $1 \le p, q \ge 2$." You might find this confusing and (unnecessarily) difficult to read. If we say "So $1 \le p$ and $q \ge 2$," the sentence is clear. It's often easier to read things if you put a word, even a little one, between symbols.
- Avoid starting a sentence with a symbol. This often confuses the reader unnecessarily. For example, consider the following sentence.

   Thus $x \in A$. A is a subset of B.

   First, the $A. A$ just doesn't look nice. Second, it's hard to read.
- Every sentence should start with a capital letter and end with a period, just like sentences are supposed to.
- All grammatical rules apply. Sentences should have nouns and verbs. Don't use the same word repeatedly.
- Strive for clarity. Always keep the reader in mind. If something follows from a definition, say so. The reader will appreciate this and will know what you are thinking *and*, what's more, you will know why what you say is true. If something follows from Theorem 10.1, say so. It is extremely important for you to be aware of when you are using a result. For one thing, it means that you are more likely to notice if you are using a result that you do not have. (This would be wrong. Don't do it.) For another, it helps the reader who may not fully understand what you are doing.
- Certain phrases are particularly helpful in guiding a reader through your proof. For example, "Suppose to the contrary, ..." tells the reader that your proof will be done by contradiction. As a second example, if you are proving "A if and only if B," your reader will understand everything better if you say, "Suppose A. ... Then we have B." And then say, "Suppose B. ... Then we have A." You should alert the reader to a proof that will be in cases, or a proof that will proceed using the contrapositive. Other examples of phrases that you may use to guide your reader will come up as we learn new techniques.

- If you can find a shorter, clearer solution, do so.
- Perhaps the most difficult thing about writing a proof is to find a balance between the main ideas in the proof and the details. You'll often find that the more you explain, the more you hide the main ideas. On the other hand, if you don't explain enough, you might overlook an important detail or confuse your reader. It's not easy to strike the right balance. This is why we suggest waiting a bit, and then rereading your proof. If you can't figure out why you did something, it's unlikely that someone else will.
- If you have a partner in the class, it is an excellent idea to exchange papers and see if things are clear to each of you. (Check with your teacher to make sure this is allowed, of course.)

Exercise 9.10.
Here's a student's proof of the following theorem: Let x and y be real numbers. Show that $xy \leq x^2/2 + y^2/2$.

Solution.

$$(x - y)^2 \geq 0$$
$$x^2 - 2xy + y^2 \geq 0$$
$$x^2 + y^2 \geq 2xy$$
$$x^2/2 + y^2/2 \geq xy$$

∎

Criticize the student's solution and rewrite the proof, paying close attention to the tips presented here. ○

For other (not necessarily independent) views on writing see [32] and [51].

10

CHAPTER

Relations

In the last chapter we introduced relations. We will now look at three useful properties of relations.

Recall that "S is a relation on a set X" is one way of saying that S is a subset of $X \times X$, and therefore the elements of S are ordered pairs, (x, y). Many authors write $x \sim y$ rather than $(x, y) \in S$. Sometimes we will write $x \sim y$ and other times we will write $(x, y) \in S$, and this is exactly the same thing. So why do it? Because sometimes one notation is more convenient than the other. Use the next exercise to familiarize yourself with both notations.

Exercise 10.1.
Let $S = \{(x, y) \in \mathbb{R} \times \mathbb{R} : x > y\}$.
  (a) Sketch the set S.
  (b) With this relation is $1 \sim 2$?
  (c) With this relation is $3.5 \sim 2$?   ◯

A relation on a set X is said to be **reflexive** if $x \sim x$ for all $x \in X$. The relation is **symmetric** if for all $x, y \in X$, whenever $x \sim y$, then $y \sim x$. Finally, the relation is **transitive** if for all $x, y, z \in X$, if $x \sim y$

and $y \sim z$, then $x \sim z$. If a relation is reflexive, symmetric, and transitive, then the relation is said to be an **equivalence relation**.

For the remainder of this chapter, we will look at examples and nonexamples of equivalence relations. There's something to note before we begin. To show that a relation is reflexive, we show that $x \sim x$ for all $x \in X$. But to show it is symmetric, we must choose two arbitrary elements of X, suppose that $x \sim y$, and then show that $y \sim x$. If we don't use the fact that $x \sim y$, we probably haven't done it correctly. Finally, to show that a relation is transitive, we must choose three arbitrary elements, suppose that $x \sim y$ and $y \sim z$, and then show that $x \sim z$. Remember to use your assumptions to show that a relation is symmetric or transitive.

Example 10.2.
Define a relation on the real numbers $\mathbb{R}$ by $x \sim y$ if and only if $x - y \in \mathbb{Z}$. Show that this relation is an equivalence relation.

Before we begin to show that this is an equivalence relation, we will do appropriate things to understand this definition. Here are a few examples of pairs that satisfy the relation:

$$3 \sim 4, \ 0 \sim -2384, \ 7 \sim 7, \ \pi \sim \pi + 7, \ -3.7 \sim 4.3.$$

On the other hand, the following pairs do not satisfy the relation:

$$3 \not\sim 3.5, \ 0 \not\sim \pi, \ -3.7 \not\sim 3.7.$$

If you have a sense of what the relation does, you are ready to move on to the proof. (If you don't have a sense of what is happening, look for more examples and nonexamples.)

Proof.
To show that this relation is reflexive, let $x \in \mathbb{R}$. Then $x - x = 0$. Since $0 \in \mathbb{Z}$, we see that $x - x \in \mathbb{Z}$. Therefore, $x \sim x$ for all $x \in \mathbb{R}$ and $\sim$ is reflexive.

To show that this relation is symmetric, let $x, y \in \mathbb{R}$. If $x \sim y$, then $x - y \in \mathbb{Z}$. But $y - x = -(x - y) \in \mathbb{Z}$, and therefore $y \sim x$. Hence this relation is symmetric.

To show that this relation is transitive, let $x, y, z \in \mathbb{R}$. If $x \sim y$ and $y \sim z$, then $x - y \in \mathbb{Z}$ and $y - z \in \mathbb{Z}$. Now the sum of two integers is an integer and therefore $x - z = (x - y) + (y - z) \in \mathbb{Z}$. In other words, $x \sim z$. Thus $x \sim z$ is transitive. ∎

In the previous example it's interesting to try to describe, in words, the set of numbers that are related to 0, 1/2, π and x. As is the case with our examples and nonexamples appearing above, we hope that describing these sets will help us to more fully understand this relation.

For 0, we look for $\{x \in \mathbb{R} : x \sim 0\} = \{x \in \mathbb{R} : x - 0 \in \mathbb{Z}\}$. Thus, the set of elements related to 0 is just $\mathbb{Z}$.

For 1/2, we look for $\{x \in \mathbb{R} : x \sim 1/2\} = \{x \in \mathbb{R} : x - 1/2 \in \mathbb{Z}\}$. Thus, the set of all elements related to 1/2 is the set $\{1/2 + k : k \in \mathbb{Z}\}$.

For π, we look for $\{x \in \mathbb{R} : x - \pi \in \mathbb{Z}\}$. Thus, the set of all elements related to π is the set $\{\pi + k : k \in \mathbb{Z}\}$.

In general it appears that for every $x \in \mathbb{R}$, the set of all elements related to x is the set $\{x + k : k \in \mathbb{Z}\}$.

Once we have an equivalence relation on a set X, we define the **equivalence class** of an element $x \in X$ to be the set E_x where $E_x = \{y \in X : x \sim y\}$. Using this notation, we see that the sets of points related to 0, 1/2, π, and x following Example 10.2 were actually descriptions of E_0, $E_{1/2}$, E_π, and E_x.

Exercise 10.3.
Write each relation below using set notation. Then decide whether or not the following relations are reflexive, symmetric, or transitive. If they are all three, prove it and describe the equivalence classes. If they are not, give a particular example to show why the property fails to hold.

  (a) Define a relation on $\mathbb{Z}$ by $x \sim y$ if and only if $x = -y$.
  (b) Define a relation on $\mathbb{Z}$ by $x \sim y$ if and only if $x - y$ is even.
  (c) Define a relation on $\mathbb{Z} \times (\mathbb{Z} \setminus \{0\})$ by $(x, y) \sim (w, z)$ if and only if $xz = yw$. ○

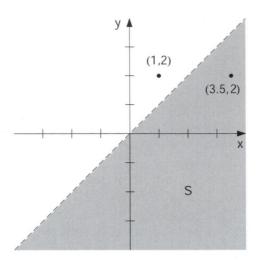

FIGURE 10.1 $\{(x,y) \in \mathbb{R} \times \mathbb{R} : x > y\}$

Solutions to Exercises

Solution to Exercise (10.1).
The set S is represented by the shaded area of Figure 10.1. Using the defined relation, $(1, 2) \notin S$ but $(3.5, 2) \in S$. Thus, the answer to (b) is no, and the answer to (c) is yes.

Solution to Exercise (10.3).
(a) This relation is neither reflexive nor transitive (but it is symmetric). It is not reflexive because, for example, $1 \neq -1$ and therefore $1 \nsim 1$. It is not transitive because $1 \sim -1$ and $-1 \sim 1$, but $1 \nsim 1$.

(b) This relation is an equivalence relation. To see this, let $x \in \mathbb{Z}$. Then $x - x = 0$ and therefore $x \sim x$, which shows that $\sim$ is reflexive. For symmetry, let $x, y \in \mathbb{Z}$. If $x \sim y$, then $x - y$ is even. Since $y - x = -(x - y)$, it follows that $y - x$ is even. Therefore $y \sim x$, and $\sim$ is symmetric. For transitivity, let $x, y, z \in \mathbb{Z}$. If $x \sim y$ and $y \sim z$, then $x - y$ and $y - z$ are both even; in other words, there exist integers m and n such that $x - y = 2m$ and $x - y = 2n$. Now, $x - z = (x-y)+(y-z) = 2m+2n = 2(m+n)$, and $m + n \in \mathbb{Z}$. Thus $x - z$ is even. Therefore $x \sim z$, and $\sim$

is transitive. We conclude that the relation is an equivalence relation.

What are the equivalence classes? You can check that if x is even, then the equivalence class corresponding to x is the set of even numbers. If x is odd, then the equivalence class corresponding to x is the set of odd numbers.

(c) This relation is an equivalence relation. It is easy to check that it is reflexive and symmetric. We check transitivity carefully here. So let (x, y), (u, v), and (w, z) be elements of $\mathbb{Z} \times (\mathbb{Z} \setminus \{0\})$. By definition, y, v, and z are all nonzero. If $(x, y) \sim (u, v)$ and $(u, v) \sim (w, z)$, then $xv = yu$ and $uz = vw$. We need to show that $xz = yw$. Multiplying both sides of the equation $xv = yu$ by z and both sides of the equation $uz = vw$ by y, we obtain the two equations $xvz = yuz$ and $uzy = vwy$. Therefore $xvz = vwy$. Now v is nonzero, so we may cancel to obtain $xz = yw$, which shows that $(x, y) \sim (w, z)$. Therefore $\sim$ is transitive.

Finally, $E_{(x,y)} = \{(w, z) \in \mathbb{Z} \times (\mathbb{Z} \setminus \{0\}) : xz = yw\}$. If we think of (x, y) as the rational number x/y, then $(w, z) \sim (x, y)$ if and only if $x/y = w/z$. So this relation is a way of identifying all fractions with the same value.

Problems

Problem 10.1.
Decide whether or not the following relations are reflexive, symmetric, or transitive. If a property holds, prove that it does. If a property does not hold, prove that it does not hold. If the relation is an equivalence relation, give the equivalence class of a general point $x \in X$.

(a) On $\mathbb{R}$, we define $x \sim y$ if and only if $x < y$.
(b) On $\mathbb{R}$, we define $x \sim y$ if and only if $x \leq y$.
(c) On $\mathbb{Z}$, we define $x \sim y$ if and only if $x - y$ is divisible by 3.
(d) On $\mathbb{R} \times \mathbb{R}$ we define $(x, y) \sim (u, v)$ if and only if $x + y = u + v$.
(e) If X is a nonempty set, define a relation on $\mathcal{P}(X)$ by $A \sim B$ if and only if $A \subseteq B$.

   (f) If X is a nonempty set, define a relation on $\mathcal{P}(X)$ by $A \sim B$ if and only if $A \setminus B \neq \emptyset$.

   (g) On $\mathbb{R} \times \mathbb{R}$, define $(x, y) \sim (w, z)$ if and only if $x = w$.

   (h) On $\mathbb{Z}$, define $x \sim y$ if and only if $|x| = |y|$.

   (i) On $\mathbb{R}^2$ define $(x, y) \sim (w, z)$ if and only if $x^2 + y^2 = w^2 + z^2$.

   (j) On $\mathbb{R}^2$ define $(x, y) \sim (w, z)$ if and only if $3x + y = 3w + z$.

Problem 10.2.

Let $X = \{1, 2, 3, 4, 5\}$.

   (a) If possible, define a relation on X that is an equivalence relation.

   (b) If possible, define a relation on X that is reflexive, but neither symmetric nor transitive.

   (c) If possible, define a relation on X that is symmetric, but neither reflexive nor transitive.

   (d) If possible, define a relation on X that is transitive, but neither reflexive nor symmetric.

Problem 10.3.

Define a relation $\sim$ on $\mathbb{R}$ as follows: For $x, y \in \mathbb{R}$, we say $x \sim y$ if and only if $x^2 - y^2 \in \mathbb{Z}$.

   (a) Prove that $\sim$ as defined above is an equivalence relation on $\mathbb{R}$.

   (b) Give five different real numbers that are in the equivalence class $E_{\sqrt{2}}$.

Problem 10.4.

Define a relation $\sim$ on $\mathbb{R}^2$ as follows: For $(x_1, x_2), (y_1, y_2) \in \mathbb{R}^2$, we say that $(x_1, x_2) \sim (y_1, y_2)$ if and only if both $x_1 - y_1$ and $x_2 - y_2$ are even integers. Is this relation an equivalence relation? Why or why not?

Problem♯ 10.5.

Let X be a nonempty set with an equivalence relation $\sim$ on it. Prove that for all elements x and y in X, the equality $E_x = E_y$ holds if and only if $x \sim y$.

Problem 10.6.
What, if anything, is wrong with the following argument? We claim that if a relation on a set X is symmetric and transitive, then it is reflexive. Here's a proof of this claim:

Proof.
Let $x \in X$. Let $y \in X$ with $x \sim y$. By symmetry we have $y \sim x$. We now use transitivity to conclude that $x \sim x$. ☐

Problem 10.7.
Give an example of a relation on $\mathbb{Z} \times \mathbb{Z}$ that is not transitive, but is reflexive and symmetric.

Problem 10.8.
Recall that a **polynomial** p over $\mathbb{R}$ is an expression of the form $p(x) = a_n x^n + a_{n-1} x^{n-1} + \cdots + a_1 x^1 + a_0$ where each $a_j \in \mathbb{R}$ and $n \in \mathbb{N}$. The largest integer j such that $a_j \neq 0$ is the **degree** of p. We define the degree of the constant polynomial $p = 0$ to be $-\infty$. (A polynomial over $\mathbb{R}$ defines a function $p : \mathbb{R} \to \mathbb{R}$.)
  (a) Define a relation on the set of polynomials by $p \sim q$ if and only if $p(0) = q(0)$. Is this an equivalence relation? If so, what is the equivalence class of the polynomial given by $p(x) = x$?
  (b) Define a relation on the set of polynomials by $p \sim q$ if and only if the degree of p is the same as the degree of q. Is this an equivalence relation? If so, what is E_r if $r(x) = 3x + 5$?
  (c) Define a relation on the set of polynomials by $p \sim q$ if and only if the degree of p is less than or equal to the degree of q. Is this an equivalence relation? If so, what is E_r, where $r(x) = x^2$?

Tips on Reading Mathematics

Don't just read it; fight it! Ask your own questions, look for your own examples, discover your own proofs. Is the hypothesis necessary? Is the converse true? What happens in the classical special case? What about the degenerate cases? Where does the proof use the hypothesis?
– Paul R. Halmos, [34]

- Be an active reader. Open to the page you need to read, get out some paper and a pencil.
- If notation is defined, make sure you know what it means. Your pencil and paper should come in handy here.
- Look up the definitions of all words that you do not understand.
- Read the statement of the theorem, corollary, lemma, or example. Can you work through the details of the proof by yourself? Try. Even if it feels like you are making no progress, you are gaining a better understanding of what you need to do.
- Once you truly understand the statement of what is to be proven, you may still have trouble reading the proof—even someone's well-written, clear, concise proof. Try to get the overall idea of what the author is doing, and then try (again) to prove it yourself.
- If a theorem is quoted in a proof and you don't know what it is, look it up. Check that the hypotheses apply, and that the conclusion is what the author claims it is.
- Don't expect to go quickly. You need to get the overall idea as well as the details. This takes time.
- If you are reading a fairly long proof, try doing it in bits.
- If you can't figure out what the author is doing, try to (if appropriate) choose a more specific case and work through the argument for that specific case.
- Draw a picture, if appropriate.
- If you really can't get it, do what comes naturally—put the book down and come back to it later. You might want to take this time to read similar proofs, some examples, or something it reminds you of.
- After reading a theorem, see if you can restate it. Make sure you know what the theorem says, what it applies to, and what it does not apply to.
- After you read the proof, try and outline the technique and main idea the author used. Try to explain it to a willing listener. If you can't do this without looking back at the proof, you probably didn't fully understand the proof. Read it again.
- Can you prove anything else using a similar proof? Does the proof remind you of something else? What are the limits of this proof? This theorem?

- If your teacher is following a book, read over the proofs before you go to class. You'll be glad you did.

As we proceed, you will have plenty of opportunities to try these tips out and find some others of your own.

11

CHAPTER

Partitions

It is sometimes helpful to split a nonempty set up into disjoint smaller pieces. For example, we might have reason to split the integers into positive integers, negative integers, and the set containing zero alone. We often split the real numbers into rational numbers and irrational numbers, or we might want to break $\mathbb{R}^2$ down into distinct vertical lines. All of these are examples of partitioning a space.

The precise definition of a partition is the following. Let X be a set. Then a family of sets $\{A_\alpha : \alpha \in I\}$ is a **partition of** X if three things happen:

(*i*) For every $\alpha \in I$, the set A_α is nonempty,
(*ii*) $\bigcup_{\alpha \in I} A_\alpha = X$, and
(*iii*) for all $\alpha, \beta \in I$, if $A_\alpha \cap A_\beta \neq \emptyset$, then $A_\alpha = A_\beta$.

Figure 11.1 provides a diagram of a partition of $X = \{a, b, c, d, e, f\}$ into sets A_1, A_2, and A_3, defined by

$$A_1 = \{a, b\}, \quad A_2 = \{c, d, e\}, \quad \text{and} \quad A_3 = \{f\}.$$

While it is often clear that the sets A_α are nonempty, you should still check. Condition (*ii*) says that every element of X is in at least one of the sets A_α. It's a sort of existence statement: for each element x of X, there exists a set A_α of which x is a member. The third

119

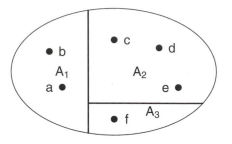

FIGURE 11.1 A partition of $\{a, b, c, d, e, f\}$.

condition is a fancy way of saying that two of the sets in the partition are either disjoint or equal. This is a sort of uniqueness statement: if x belongs to two sets A_α and A_β, then the sets must be equal. We hope the examples below will clarify these concepts.

Example 11.1.
For each $x \in \mathbb{R}$, define $A_x = \{y \in \mathbb{R} : |x| = |y|\}$. We will show that $\{A_x : x \in \mathbb{R}\}$ forms a partition of $\mathbb{R}$.

Before we begin, note that the sets A_α are not mutually disjoint. For example, $A_1 = \{1, -1\}$ and $A_{-1} = \{-1, 1\}$, so $A_1 \cap A_{-1} \neq \emptyset$. That will not be a problem though, and it does illustrate why we stated condition (*iii*) the way we did. It happens that $A_1 \cap A_{-1} \neq \emptyset$, but $A_1 = A_{-1}$, as required.

Proof.
We note that all the sets A_α are nonempty. So there are two things left to show, namely conditions (*ii*) and (*iii*) of the definition of partition. We begin by showing that (*ii*) holds; that is $\bigcup_{x \in \mathbb{R}} A_x = \mathbb{R}$. First, $\bigcup_{x \in \mathbb{R}} A_x \subseteq \mathbb{R}$ since each $A_x \subseteq \mathbb{R}$. We show the reverse containment using an element-chasing argument. Let $y \in \mathbb{R}$. Then $|y| = |y|$, so $y \in A_y$. Since $y \in A_x$ for some x (namely $x = y$), we may conclude that $y \in \bigcup_{x \in \mathbb{R}} A_x$. Thus $\bigcup_{x \in \mathbb{R}} A_x = \mathbb{R}$.

Next, suppose that $A_x \cap A_y \neq \emptyset$. Then we must show that $A_x = A_y$. By our assumption, there exists $z \in \mathbb{R}$ such that $z \in A_x \cap A_y$. Therefore, $|x| = |z|$ and $|y| = |z|$. In particular, $|x| = |y|$. So,

$$A_x = \{w \in \mathbb{R} : |x| = |w|\} = \{w \in \mathbb{R} : |y| = |w|\} = A_y.$$

∎

Well, we showed the two sets A_x and A_y are equal with nary an element-chasing argument in sight. What happened? We certainly could have started with an element from one side and showed it was in the other, switched sides, repeated what we did and then concluded we were done. But this is somewhat cumbersome and doesn't show us what is really going on. So from now on, even though we can use element-chasing, we are going to use whatever produces the most elegant or enlightening proof.

Exercise 11.2.
For each $n \in \mathbb{N}$, let $A_n = [-n, n]$. Show that $\{A_n : n \in \mathbb{N}\}$ does not form a partition of $\mathbb{R}$. However, if we define $B_n = [n, n+1)$, then $\{B_n : n \in \mathbb{Z}\}$ does partition $\mathbb{R}$. $\bigcirc$

In Example 11.1 and Exercise 11.2, the third condition may have reminded you of transitivity. If so, then it may not surprise you to learn that there is a connection between equivalence relations and partitions. As we shall see, every equivalence relation on a set X gives rise to equivalence classes in a natural way. These equivalence classes are sets and these sets partition our set X.

Conversely, a family of sets that partitions a set X gives rise to an equivalence relation on X. How? Well, we say two elements in X are related if they belong to the same set of the partition. We shall now show that this relation is an equivalence relation. We can shorten our proof of this theorem, if we first prove something less ambitious. A helpful result that is used to prove a theorem is called a lemma. Lemmas are sometimes of independent interest.

Lemma 11.3.
Let X be a set and let $\sim$ be an equivalence relation on X. For two arbitrary elements x and y in X, if $E_x \cap E_y \neq \emptyset$, then $E_x = E_y$.

The very first thing we should probably ask ourselves before beginning our proof is: What is E_x? If we don't know, we can't understand the proof. So, before reading the proof, we write down the definition:

$$E_x = \{z \in X : x \sim z\}.$$

Now the proof should be easy.

Proof.
Let $z \in E_x$. Hence $x \sim z$. Since we assume that $E_x \cap E_y \neq \emptyset$, we may choose $w \in E_x \cap E_y$. Thus $w \in E_x$, and therefore $x \sim w$. Similarly, $w \in E_y$ and therefore $y \sim w$. By symmetry, $w \sim x$. So $y \sim w$, $w \sim x$, and $x \sim z$. By transitivity, $y \sim z$. Thus, $z \in E_y$, and we may conclude that $E_x \subseteq E_y$.

Exactly the same argument $(\star)$ shows that $E_y \subseteq E_x$. Hence $E_x = E_y$. ∎

One comment on the proof above: When we use the words "exactly the same argument," as in $(\star)$, that means nothing would be changed except (possibly) the symbols. If you use words to that effect (like "similarly" or "exactly as above"), make sure that what you say is true. Now that we have our lemma, we turn to the proof of our main theorem.

Theorem 11.4.
Let $\sim$ be an equivalence relation on a nonempty set X. Then the family of equivalence classes $\{E_x : x \in X\}$ is a partition of X. Furthermore, if $\{A_\alpha : \alpha \in I\}$ is a partition of a set X and we define $x \sim y$ if and only if $x, y \in A_\alpha$ for some $\alpha \in I$, then $\sim$ is an equivalence relation on X.

Before beginning the proof, let's reflect on what we need to do. For the first assertion ("$\{E_x : x \in X\}$ is a partition") we need to show that the sets are nonempty, and satisfy conditions *(ii)* and *(iii)* in the definition of partition.

What do we expect to use? Our assumptions, of course. We are assuming $\sim$ is an equivalence relation, so we should use the fact that $\sim$ is reflexive, symmetric, and transitive. But that's only one direction—this would show that an equivalence relation gives rise to equivalence classes and these, in turn, form a partition of our set.

For the other direction, we want to show that if we have a relation defined by a partition, then the relation is an equivalence relation. So that means we must show that $\sim$ is reflexive, symmetric, and transitive. How will we do that? Well, probably the first thing to do is to make sure we know what $\sim$ is. Remember, $x \sim y$ if and only if there exists α such that $x, y \in A_\alpha$. Now, finally, we may begin.

Proof.
First we'll show that given an equivalence relation on X, the family of sets $\{E_x : x \in X\}$ forms a partition of X. We first show that each E_x is nonempty. Since the relation is reflexive, $x \sim x$ for each $x \in X$. Thus $x \in E_x$ for each $x \in X$, and $E_x \neq \emptyset$.

Now we need to check condition (ii): that $\bigcup_{y \in X} E_y = X$. If $x \in X$, then we have just seen that $x \in E_x$. This shows that $x \in \bigcup_{y \in X} E_y$. Thus $X \subseteq \bigcup_{y \in X} E_y$. Since the opposite inclusion follows from the fact that $E_y \subseteq X$ for each $y \in X$, we know that $X = \bigcup_{y \in X} E_y$. Thus, condition (ii) holds.

To show that condition (iii) holds, suppose that for $x, y \in X$, we have $E_x \cap E_y \neq \emptyset$. By Lemma 11.3, we conclude that $E_x = E_y$, and condition (iii) holds. Thus, the set of equivalence classes $\{E_x : x \in X\}$ satisfies conditions (i), (ii), and (iii) and therefore forms a partition of X.

For the converse, suppose that $\{A_\alpha : \alpha \in I\}$ forms a partition of X. By condition (ii), $X = \bigcup_{\alpha \in I} A_\alpha$. Thus, for $x \in X$, there exists $\alpha \in I$ such that $x \in A_\alpha$. Since x is in the same set as itself, $x \sim x$. Since x was arbitrary, $\sim$ is reflexive.

Suppose now that $x, y \in X$ and $x \sim y$. Then there exists $\alpha \in I$ such that $x, y \in A_\alpha$. But if $x, y \in A_\alpha$, then $y, x \in A_\alpha$. Consequently, $y \sim x$. Therefore $\sim$ is symmetric.

Finally, suppose that $x, y, z \in X$ where $x \sim y$ and $y \sim z$. We must show that $x \sim z$. By the definition of $\sim$ we see that there exists $\alpha \in I$ such that $x, y \in A_\alpha$, and there exists $\beta \in I$ such that $y, z \in A_\beta$. Therefore, $A_\alpha \cap A_\beta \neq \emptyset$. By property (iii) of partitions, $A_\alpha = A_\beta$. Thus $x, z \in A_\alpha$. Therefore, $x \sim z$, as desired. We conclude that the partition gives rise to an equivalence relation, since $\sim$ is symmetric, transitive, and reflexive. ∎

Exercise 11.5.
For $r \in \mathbb{R}$, let $A_r = \{(x, y) \in \mathbb{R}^2 : x + y = r\}$. Show that $\{A_r : r \in \mathbb{R}\}$ is a partition of $\mathbb{R}^2$. Then describe the equivalence relation and equivalence classes associated with this partition. ○

Solutions to Exercises

Solution to Exercise (11.2).
The family $\{A_n : n \in \mathbb{Z}\}$ does not partition $\mathbb{R}$ because condition *(iii)* is not satisfied: $A_1 \cap A_2 \neq \emptyset$, but $A_1 \neq A_2$. The family $\{B_n : n \in \mathbb{Z}\}$ does partition $\mathbb{R}$: For each $n \in \mathbb{Z}$, the set B_n is nonempty, the union of the sets satisfies

$$\bigcup_{n \in \mathbb{Z}} B_n = \bigcup_{n \in \mathbb{Z}} [n, n+1) = \mathbb{R},$$

and if $B_n \cap B_m \neq \emptyset$, then $[n, n+1) \cap [m, m+1) \neq \emptyset$. Since m and n are integers, the intervals $[n, n+1)$ and $[m, m+1)$ are either equal or disjoint. We conclude that $[n, n+1) = [m, m+1)$; in other words, $B_n = B_m$.

Solution to Exercise (11.5).
Note that for $r \in \mathbb{R}$, the ordered pair $(0, r)$ satisfies the condition $0 + r = r$. Thus $(0, r) \in A_r$ and A_r is nonempty. Since it is clear that $\bigcup_{r \in \mathbb{R}} A_r \subseteq \mathbb{R}^2$, we check the reverse inclusion. So let $(u, v) \in \mathbb{R}^2$. Then $s = u + v \in \mathbb{R}$ and consequently $(u, v) \in A_s$. Thus $(u, v) \in \bigcup_{r \in \mathbb{R}} A_r$, and $\bigcup_{r \in \mathbb{R}} A_r = \mathbb{R}^2$, completing the proof of condition *(ii)* in the definition of partition. Finally, suppose that $A_r \cap A_s \neq \emptyset$. Then there exists $(u, v) \in A_r \cap A_s$. By the definition of A_r and A_s this means that $r = u + v = s$. Thus, $A_r = A_s$, as desired.

The associated equivalence relation on $\mathbb{R}^2$ is defined as follows. For $(x, y), (u, v) \in \mathbb{R}^2$, we will say $(x, y) \sim (u, v)$ if and only if $x + y = u + v$. By our work above and Theorem 11.4, this is an equivalence relation on $\mathbb{R}^2$. The equivalence classes are the lines with slope -1.

In the two exercises in this chapter, the third condition (of partition) is satisfied because the indices (n and m in Exercise 11.2, and r and s in Exercise 11.5) are equal. Though this can happen, Example 11.1 shows that the two sets can be equal without the indices being equal. Condition *(iii)* in the definition of partition requires that we show that the two sets are equal—not the two indices.

Problems

Problem 11.1.
For each of the relations in Problem 10.1 that you determined to be equivalence relations, describe the partition associated with it.

Problem 11.2.
Determine whether or not the following are equivalence relations on $\mathbb{R}^2$. If they are, describe the partition associated with each:
(a) $(x, y) \sim (w, z)$ if and only if $y = w$;
(b) $(x, y) \sim (w, z)$ if and only if $x^2 = w^2$;
(c) $(x, y) \sim (w, z)$ if and only if $xw = yz$.

Problem 11.3.
(a) For each $r \in \mathbb{R}$, let $A_r = \{(x, y, z) \in \mathbb{R}^3 : x + y + z = r\}$. Is this a partition of $\mathbb{R}^3$? If so, give a geometric description of the partitioning sets.
(b) For each $r \in \mathbb{R}$, let $A_r = \{(x, y, z) \in \mathbb{R}^3 : x^2 + y^2 + z^2 = r^2\}$. Is this a partition of $\mathbb{R}^3$? If so, give a geometric description of the partitioning sets.

Problem 11.4.
(a) Let $A = \{1, 2, \ldots, 10\}$. Describe a partition of A that gives rise to five distinct partitioning sets.
(b) Describe a partition of $\mathbb{Z}$ that gives rise to five distinct partitioning sets.
(c) Can you describe a partition of $\mathbb{R}$ that gives rise to five distinct partitioning sets?

Problem 11.5.
(a) Suppose that we partition $\mathbb{R}^3$ into horizontal planes. What equivalence relation is associated with this partition?
(b) Suppose that we partition $\mathbb{R}^3$ into concentric spheres, centered at $(0, 0, 0)$. What equivalence relation is associated with this partition?

Problem 11.6.
Suppose that we look at the set X containing all circles in the plane. Define an equivalence relation on this set of circles by $c \sim d$ if and only if the circles c and d have the same center. Describe the partition associated with this equivalence relation.

Problem 11.7.
Consider the set P of polynomials with real coefficients. Decide whether or not each of the following determine a partition of P. If you decide that it does determine a partition, show it carefully. If you decide that it does not determine a partition, justify your answer. (See Problem 10.8 for more information about polynomials.)
  (a) For $m \in \mathbb{N}$, let A_m denote the set of polynomials of degree m.
  (b) For $c \in \mathbb{R}$, let A_c denote the set of polynomials such that $p(0) = c$.
  (c) For a polynomial q, let A_q denote the set of all polynomials p such that q is a factor of p; that is, there is a polynomial r such that $p = qr$.
  (d) For $c \in \mathbb{R}$, let A_c denote the set of polynomials such that $p(c) = 0$.

Problem 11.8.
For two nonempty disjoint sets, I and J, let $\{A_\alpha : \alpha \in I\}$ be a partition of $\mathbb{R}^+$ and $\{A_\alpha : \alpha \in J\}$ be a partition of $\mathbb{R}^- \cup \{0\}$. Prove that $\{A_\alpha : \alpha \in I \cup J\}$ is a partition of $\mathbb{R}$.

Problem 11.9.
Let X be a nonempty set and $\{A_\alpha : \alpha \in I\}$ be a partition of X.
  (a) Let B be a subset of X such that $A_\alpha \cap B \neq \emptyset$ for every $\alpha \in I$. Is $\{A_\alpha \cap B : \alpha \in I\}$ a partition of B? Prove it or give a counterexample.
  (b) Suppose further that $A_\alpha \neq X$ for every $\alpha \in I$. Is $\{X \setminus A_\alpha : \alpha \in I\}$ a partition of X? Prove it or give a counterexample.

Problem 11.10.
Recall that for an integer n, the symbol $3|n$ means that there exists $m \in \mathbb{Z}$ such that $n = 3m$. For each integer i, where $i = 0, 1, 2$, we

define $A_i = \{x \in \mathbb{Z} : 3|(x-i)\}$. Show that $\{A_i : i = 0, 1, 2\}$ is a partition of $\mathbb{Z}$.

Tips on Putting It All Together

Now we will build upon the foundations we have created.

- In each section, work through the definitions. (Check "Tips on Definitions.") If you don't know the definitions, you cannot get started. So the first step is to make sure that you have *mastered* them.
- Next, learn and understand all theorems. You don't have to memorize their number, of course, but you should know by name each theorem that has a name. Make sure you can restate every theorem in the text correctly.
- If you are asked to prove something, look for a proof or theorem that reminds you of your problem. Read it over.
- If your problem is too difficult, try a simpler one first.
- Whenever you claim something is true, say why (at least to yourself, if it is minor and to the reader, if it is major). Is it a definition? a theorem? Are the techniques the same as in a proof everyone has already seen? If you are writing up your homework, tell the grader which theorem (now you should give a number) or what definition you are using.
- If you can check your solution, do so. Is your answer reasonable?
- Does your theorem make sense? Does it agree with other theorems in the text? Does your proof use everything you were given?
- Your first draft is precisely that. No one should have to read someone else's first draft. Work out the solution, write it up, put it away, read it again, and rewrite it.

12

Order in the Reals

You've seen numbers ever since you've been in school, and you know a lot about them. It is possible to give them a careful mathematical foundation. In fact, it's possible to construct the natural numbers (and you can do so in Project 27.3). Then, if you try to introduce operations like addition and subtraction, you'll find that you are missing something: the negative numbers. So you look at the integers, and try again. Now, trying to introduce multiplication and division, you'll find you are missing something again: multiplicative inverses. So you look at the rational numbers, and you'll find you are missing something yet again. That brings you to the real numbers.

In this chapter, we will assume that we have the real numbers and that they satisfy the algebraic axioms listed in the appendix. We will also assume that $\mathbb{R}$ has the order relation $\leq$ satisfying the other properties listed in the appendix. It is our goal here to discuss what's missing in $\mathbb{Q}$, and show you why $\mathbb{R}$ has what's missing. This is known as completeness of $\mathbb{R}$.

What do we mean by this property "completeness," that $\mathbb{R}$ has, but $\mathbb{Q}$ doesn't? If we take a stroll along the real number line, we can walk right up to any real number and it will be there waiting for us. In contrast to this, if we walk along the real line, this time stepping on rational numbers only, we might walk right up to where we'd expect

$\sqrt{2}$ to be, but it will be out having lunch with other irrationals. We'll make this precise by the end of this chapter. We now turn to the important terminology that we need.

A nonempty subset A of $\mathbb{R}$ is **bounded above** if there is a real number M such that $x \leq M$ for all $x \in A$. A real number M satisfying $x \leq M$ for all $x \in A$ is called an **upper bound** of A. The nonempty subset A of $\mathbb{R}$ is **bounded below** if there is a real number m such that $m \leq x$ for all $x \in A$. A real number m satisfying $m \leq x$ for all $x \in A$ is called a **lower bound** of A. We say a nonempty set is **bounded** if it is bounded above and below. For example, the open interval $(0, 1)$ is bounded above, since every $x \in (0, 1)$ satisfies $x \leq 1$. The number 1 is an upper bound, and so is the real number 1.5. In fact, every number greater than 1 is an upper bound. Similarly, since $x \geq 0$ for all $x \in (0, 1)$, the set $(0, 1)$ is bounded below and 0 is an example of a lower bound of the set. Since $(0, 1)$ is bounded above and below, it is an example of a bounded set.

Exercise 12.1.
For each of the following sets of real numbers, decide whether it is bounded above, bounded below, and (consequently) whether or not it is bounded. If the set is bounded above, give three different examples of upper bounds in $\mathbb{R}$. If the set is bounded below, give three different examples of lower bounds in $\mathbb{R}$. Use your intuition; we'll prove things rigorously later. The sets are:
  (a) $\{x \in \mathbb{R} : x^2 \leq 5\}$;
  (b) $\{x \in \mathbb{R} : x^3 < 5\}$;
  (c) $\{x \in \mathbb{N} : x \leq 5\}$;
  (d) $\{x \in \mathbb{Q} : x^2 < 2\}$. ○

Sometimes a subset of $\mathbb{R}$ that is bounded above contains a largest element, and we give this element a special name: a maximum. The real number M is a **maximum** of the set A, if $M \in A$ and $x \leq M$ for all $x \in A$. We will write $M = \max A$ for a maximum of the set A. Note that a maximum is an upper bound that lies in the set A. Likewise, a real number m is a **minimum** of the set A, if $m \in A$ and $m \leq x$ for all x in A. We will write $m = \min A$ to denote a minimum of the set

A. Again, notice that a minimum is a lower bound that lies in the set A.

It should now be clear that if a set A has a maximum, then A must be bounded above, and if the set has a minimum, then A must be bounded below. What about the converse?

Example 12.2.

Give an example of a bounded set that has neither a maximum nor a minimum.

We claim that the set $(0, 2)$ is bounded and has neither a maximum nor a minimum.

Proof.

For each $x \in (0, 2)$, we know that $0 < x < 2$. Therefore 0 is a lower bound of the set and 2 is an upper bound. Thus, $(0, 2)$ is bounded. To see that it has no maximum, suppose to the contrary that s is a maximum of the set $(0, 2)$. Then, by definition of maximum, s must be in the set, so $0 < s < 2$. But (as you can check) $0 < s < (2 + s)/2 < 2$, and therefore $(2 + s)/2$ is in the set $(0, 2)$ and larger than s, a contradiction. In a similar fashion, you can check that there is no minimum. ∎

It turns out that there is an upper bound that can help us when we don't have a maximum (called the supremum), and a lower bound that can help us when we don't have a minimum (called the infimum). We define these below.

Let A be a nonempty set of real numbers that is bounded above. Then a real number U is said to be a **supremum** of A or **least upper bound** of A if

(i) $a \leq U$ for all $a \in A$, and
(ii) if $M \in \mathbb{R}$ satisfies $a \leq M$ for all $a \in A$, then $U \leq M$.

Note that (i) says that U is an upper bound, while (ii) says that U is least among all upper bounds. While the phrase "least upper bound" is more descriptive, most authors prefer the term "supremum."

The following lemma tells us that the supremum, when it exists, is unique.

Lemma 12.3.
If a nonempty subset of $\mathbb{R}$ *has a supremum, then the supremum is unique.*

We first try to understand the problem. Let's call our set S. We are not asking whether or not a supremum of S exists. What we are trying to do is to show that there cannot exist two different real numbers a and b, such that both a and b fulfill the properties of supremum of S. (We've seen examples of sets with more than one upper bound. Maybe there are sets with more than one least upper bound.)

So we turn to devising a plan. Let's suppose that there are two such numbers a and b, and try to show that they must be equal.

Proof.
Let S be a nonempty subset of $\mathbb{R}$. Suppose a and b are two real numbers that satisfy properties (i) and (ii) in the definition of supremum. Then a is an upper bound. Since b is a supremum, property (ii) implies that $b \leq a$. On the other hand, since b is an upper bound and a is a supremum, property (ii) implies that $a \leq b$. Thus $a = b$, and we conclude that there is at most one supremum. ∎

From here on in, we will refer to "the" supremum, and we will denote the supremum of a nonempty set A by $\sup A$.

The last proof was your first proof of uniqueness. This particular proof is fairly standard. You'll frequently be able to prove uniqueness by supposing that you have two such objects, and showing that they must be equal.

Exercise 12.4.
Return to Exercise 12.1. Use your intuition to decide which of the sets have a supremum in $\mathbb{R}$. For the sets that you decide have a supremum, find a real number that you believe is the supremum. ○

So we have defined two notions—supremum and maximum. What are the differences and what are the similarities? A close look at the definition shows that the maximum of a set A must be in A, while the supremum of A need not. On the other hand, we also have the following.

Exercise 12.5.
Let A be a nonempty subset of real numbers that is bounded above. Show that if A has a maximum M, then M is the supremum of A. Conclude that a maximum of a set is unique and thus we can speak of "the" maximum of a set A—if it exists. ○

In Exercise 12.6 you will define and investigate the infimum (or greatest lower bound) of a set. This is an important exercise, and we will refer to it frequently.

Exercise 12.6.
Let A be a nonempty subset of $\mathbb{R}$ that is bounded below.
  (a) Define **lower bound** and **infimum** (or **greatest lower bound**) of the set A.
  (b) Do what you always do when confronted with a new definition: find examples and nonexamples.

The infimum of a set A, denoted $\inf A$, is also unique (Problem 12.14) and hence we can speak of "the" infimum. If a set A has a minimum, then this minimum is the infimum of the set. Thus, $\min A$ is also unique, if it exists.

Some students find the words supremum and infimum difficult to remember. But once you get used to it, these words will sound like what they are: If the supremum is in the set, it's the maximum. If it's not in the set, the supremum (as suggested by the word "superior") lies above the set. Similarly, if the infimum is in the set, it's the minimum. If it's not in the set, then the infimum (as suggested by the word "inferior") lies below the set.

The next example shows how to rigorously prove that a particular number is the infimum (or supremum) of a set. Remember that to show l is the infimum, we must show that it is a lower bound, and that if y is another lower bound, then $y \leq l$. We will actually show the contrapositive of this last assertion: if $y > l$, then y is not a lower bound.

Example 12.7.
Show that $\inf(3, 4] = 3$.

Proof.

We note that $3 \le x$ for all $x \in (3, 4]$. Therefore, 3 is a lower bound.

To see that it is the infimum, we will show that nothing larger can be a lower bound. To this end, let y be chosen so that $3 < y$. If $y > 4$, then y is not a lower bound of $(3, 4]$. If $y \le 4$, then $(3+y)/2$ is a real number such that $3 < (3+y)/2 < y \le 4$. Therefore $(3+y)/2 \in (3, 4]$ and $(3+y)/2 < y$. Thus, y is not a lower bound of $(3, 4]$. Hence, 3 is the infimum of $(3, 4]$. ∎

Here we have an example of a bounded set that has no minimum, but it does have an infimum.

We saw in Example 12.2 that there are sets that are bounded, but have no maximum or minimum. Since a maximum is a supremum, it may seem that there exist sets that are bounded above but have no supremum in $\mathbb{R}$. It turns out that in $\mathbb{R}$ this is not the case; the real numbers are constructed to guarantee the existence of a supremum of every bounded nonempty set. This will not be proved; in a way it is an agreement. The technical term for such a statement is axiom.

The completeness axiom of $\mathbb{R}$.

Every nonempty subset of real numbers that is bounded above has a supremum.

Exercise 12.8.

State a version of the completeness axiom of $\mathbb{R}$ replacing the word "supremum" by the word "infimum." What conditions, if any, must be placed on the set? ○

Here's an extremely useful consequence of the work we have built up in this chapter:

Theorem 12.9 (Archimedean property of $\mathbb{R}$).

Let a and b be two positive real numbers. Then there exists a positive integer n such that $a < nb$.

Proof.

Suppose that this is not true; that is, suppose that there are two positive real numbers, a and b such that $a \ge nb$ for all $n \in \mathbb{N}$. This means that a/b is an upper bound of $\mathbb{N}$. Therefore, by the completeness axiom of $\mathbb{R}$, it follows that $\mathbb{N}$ has a supremum, which we will call u. Now consider $u - 1$. Since this is less than the supremum u, it can't be an upper bound of $\mathbb{N}$. So there exists $m \in \mathbb{N}$ with $m > u - 1$. Therefore, $m + 1 \in \mathbb{N}$ and $m + 1 > u$. Since no element of $\mathbb{N}$ can be greater than the upper bound u, this is a contradiction. ∎

A useful special case occurs when we let $b = 1$ in the statement of the Archimedean property. When we have a result that follows from a theorem that we just proved, we call it a corollary. Thus we have

Corollary 12.10.
For every real number a, there is an integer n such that $a < n$.

Proof.
If $a \leq 0$, the result is obvious. If $a > 0$, this follows from the Archimedean property of $\mathbb{R}$. ∎

We now turn to the well-ordering principle of $\mathbb{N}$, which is concerned with a fundamental property of the natural numbers. There is another important principle, called the principle of mathematical induction, which we will introduce in Chapter 17 and Project 27.3. If you work the project, you will learn that induction is one of the five Peano axioms that can be used to construct the natural numbers. For now, we will state and use the well-ordering principle without proof. In Chapter 17, we will show that the well-ordering principle of $\mathbb{N}$ and the principle of mathematical induction are equivalent.

Well-ordering principle of $\mathbb{N}$.
Every nonempty subset of the natural numbers contains a minimum.

As a consequence of the well-ordering principle, we obtain an interesting theorem about where the rationals "live." The next theorem suggests that they can really fill up space! The curious thing about them, which we will return to in Chapter 22, is that there really aren't that many of them.

Theorem 12.11.
Let a and b be two real numbers satisfying a < b. Then there is a rational number c such that a < c < b.

The proof of this result will be much easier to follow if you understand the basic idea. It's this: if the difference between a and b were greater than one, then there would have to be an integer m with $a < m < b$ and we would be done. Of course, the difference does not have to be greater than one, but we can sort of force it to be: Look at $b - a$ and multiply by an integer n so that $n(b - a) > 1$. Now the difference between nb and na is greater than one, so there has to be an integer m between them (but this needs proof). So we will prove that there exists an integer m with $na < m < nb$. Divide by n to obtain the desired rational number, m/n.

Proof.
Note that we may assume, without loss of generality, that $a > 0$. (Why?) By Theorem 12.9 there is an integer n such that $n(b-a) > 1$. Thus,

$$nb > 1 + na. \tag{12.1}$$

Now consider the subset A of $\mathbb{N}$ defined by $A = \{r \in \mathbb{N} : na < r\}$. By Corollary 12.10, A is nonempty. The well-ordering principle implies that A has a minimum, which we call m. Thus $m \in A$, and from the definition of A we see that $na < m$; in other words, $a < m/n$. Let c be the rational number m/n. Then we have the lower inequality, $a < c$, and we are halfway there. For the upper inequality, note that $m - 1$ is not in the set A (what would happen if it were?) so $na \geq m - 1$. So, putting this together with equation 12.1 we get

$$nb > 1 + na \geq 1 + (m - 1).$$

So $nb > m$, and $b > m/n$. Now $c = m/n$ is a rational number between a and b, and this completes the proof. ∎

One interesting consequence of the completeness axiom is that we can now give a rigorous proof that if a is a positive real number, then a has a positive square root; that is, there exists $x \in \mathbb{R}^+$ such that $x^2 = a$. We'll do the proof for $a = 2$, and you can easily modify it to hold for all other positive real numbers. In the rest of the text, we'll assume the usual rules for exponents and roots of real numbers.

Theorem 12.12 (Existence of square roots in $\mathbb{R}^+$).
There exists a positive real number a such that $a^2 = 2$.

In order to help you better understand this proof, we'll indicate how we "devised a plan." The basic idea is that the square root should be the supremum of the set $A = \{x \in \mathbb{R}^+ : x^2 < 2\}$. The completeness axiom tells us this supremum exists, so we'll call it a. How do we show that a is the square root of 2? Well, we need to show that $a^2 = 2$. If $a^2 > 2$, our intuition tells us that we should be able to subtract something off of a, which we'll call x, and come up with something smaller than a that is still an upper bound of the set. Thus $a - x$ would be an upper bound smaller than the supremum, and this would contradict the fact that a is the least upper bound. So we now have to worry about the case in which $a^2 < 2$. In this case, our intuition tells us that we should be able to add just a little bit to a, which we'll call y, and find an element of A, namely $a + y$, that is bigger than the upper bound a. This now contradicts the fact that a is an upper bound. So, since neither of these two cases can occur, the only other possibility, $a^2 = 2$, must be the one that holds.

Proof.
Let $A = \{x \in \mathbb{R}^+ : x^2 < 2\}$. Then A is a nonempty subset of $\mathbb{R}$ (since $1 \in A$), and A is bounded above (by, for example, 2). By the completeness axiom A has a supremum, which we denote by a. What we know so far is that a is a real number, and $1 \leq a \leq 2$. We will show that $a^2 = 2$.

Before we show that $a^2 = 2$, we claim that the following is true:

$$\text{If } 0 < c < a, \text{ then } c^2 < 2. \tag{12.2}$$

To establish our claim, let c be chosen with $0 < c < a$. Then c is less than the supremum of A, so c cannot be an upper bound of A. Thus, there exists $b \in A$ with $c < b$. Since $b \in A$, we know that $b^2 < 2$. Therefore, $c^2 < b^2 < 2$, establishing the claim.

We know that one of the following three cases must occur: $a^2 > 2$, $a^2 < 2$, or $a^2 = 2$. We'll show that the first two cases are impossible.

Case 1. Suppose that $a^2 > 2$. Let $x = (a^2 - 2)/(2a)$. Then, as you can check, $0 < a - x < a$. By our claim (equation 12.2), $(a-x)^2 < 2$. But

$$(a - x)^2 = a^2 - 2ax + x^2 \geq a^2 - 2ax = 2.$$

So $(a-x)^2 < 2$, and $(a-x)^2 \geq 2$. This is impossible and we conclude that case 1 cannot occur.

Case 2. Suppose that $a^2 < 2$. Let $y = (2 - a^2)/(3a)$. Then $y > 0$, and therefore $a + y > a$. Since a is an upper bound of A, we see that $a + y \notin A$. Thus $(a + y)^2 \geq 2$. We will show that $y < a$, and then use this to obtain our contradiction. To this end, note that because $1 \leq a$ we have

$$y - a = (2 - 4a^2)/(3a) < 0.$$

Thus, $y < a$, as claimed. Since $y < a$, we also know that $y^2 < ay$. Thus,

$$(a + y)^2 = a^2 + 2ay + y^2 < a^2 + 2ay + ay = a^2 + 3ay = 2.$$

So, $(a+y)^2 \geq 2$, and $(a+y)^2 < 2$. This is impossible, and we conclude that case 2 cannot occur.

Thus we conclude that the only remaining possibility holds, and therefore $a^2 = 2$. ∎

The completeness axiom of $\mathbb{R}$ says that if we start with a nonempty set that is bounded above, we can find its supremum. In $\mathbb{Q}$, this is not the case: We can find a nonempty subset of $\mathbb{Q}$ that is bounded above but has no supremum in $\mathbb{Q}$. In other words, there is no completeness axiom for $\mathbb{Q}$.

Example 12.13.
Show that the set $\mathbb{Q}$ is not complete; that is, show that there is a nonempty set B of rational numbers that is bounded above, but no rational number b satisfies both
  (i) $x \leq b$ for all $x \in B$, and
  (ii) if $c \in \mathbb{Q}$ and $x \leq c$ for all $x \in B$, then $b \leq c$.

Proof.
Let $B = \{x \in \mathbb{Q}^+ : x^2 < 2\}$. Then B is nonempty ($1 \in B$) and bounded above (2 is an upper bound). Suppose to the contrary that the rational number b satisfies conditions (i) and (ii) above. If such a rational number b exists, it must satisfy one of the following three things: $b = \sqrt{2}$, $b > \sqrt{2}$, or $b < \sqrt{2}$. We'll show that no rational number can satisfy one of these.

We know from Theorem 5.2 that $\sqrt{2}$ is not rational, and we know from our assumptions that b is rational. So $b \neq \sqrt{2}$.

Suppose that $b > \sqrt{2}$. By Theorem 12.11, there is a rational number c such that $\sqrt{2} < c < b$. Now the supremum of the set $A = \{x \in \mathbb{R}^+ : x^2 < 2\}$ is, as we have just seen in Theorem 12.12, $\sqrt{2}$. For every $x \in B$ we know that $x \in A$ and consequently $x \leq \sqrt{2} < c$. Thus c satisfies the hypotheses of (ii). But $c < b$, and therefore c does not satisfy the conclusion of (ii). This implies that this case cannot occur.

Now suppose that $b < \sqrt{2}$. By Theorem 12.11, there is a rational number c with $b < c < \sqrt{2}$. The right side of this inequality tells us that $c^2 < 2$ and therefore $c \in B$. But $c > b$, and this contradicts the fact that b satisfies condition (i) above. This implies that this case cannot occur.

Thus we conclude that there is no rational number satisfying conditions (i) and (ii) above. ■

By the way, there is still something "missing" in $\mathbb{R}$—the square root of -1. So you might decide to look at complex numbers ... but that's another story.

Solutions to Exercises

Solution to Exercise (12.1).
We include brief answers to each part here.
- (a) This set is bounded, and therefore bounded above and below. Some possible upper bounds are $\sqrt{5}$, 3, and 121. Some possible lower bounds are $-\sqrt{5}$, -10, and -2π.

(b) This set is bounded above, and it is not bounded below. Some possible upper bounds are $5^{1/3}$, 10, and 21.3.
(c) This set is bounded, and therefore bounded above and below. Some possible upper bounds are 5, 121, and 1000. Some possible lower bounds are 0, −3, and −12.
(d) This set is bounded above and below, and therefore bounded. Every real number greater than or equal to $\sqrt{2}$ will work as an upper bound, and every real number less than or equal to $-\sqrt{2}$ will work as a lower bound.

Solution to Exercise (12.5).
Let $M = \max A$. Then $a \leq M$ for all $a \in A$ and property (i) of the definition of supremum is fulfilled. Now suppose that K is a real number satisfying $a \leq K$ for all $a \in A$. Since M is in A we have in particular that $M \leq K$. Thus, property (ii) holds and $M = \sup A$.

By Lemma 12.3, the supremum of A is unique. Since $\max A$ is the supremum of A, it is also unique.

Solution to Exercise (12.6).
Let A be a nonempty set of real numbers that is bounded below.
(a) A number m is a lower bound of A if $a \geq m$ for all $a \in A$.
(b) A number m is the infimum (or greatest lower bound) of A if
   (i) $a \geq m$ for all $a \in A$, and
   (ii) if y is a real number satisfying $a \geq y$ for all $a \in A$, then $m \geq y$.

Solution to Exercise (12.8).
Completeness axiom of $\mathbb{R}$; infimum version. *Every nonempty subset of real numbers that is bounded below has an infimum.*
   We note that the infimum version follows from the completeness axiom stated in this chapter. To see this, you would complete the following outline of a proof: If you have a nonempty subset S that is bounded below, the subset $-S = \{-x : x \in S\}$ is nonempty and bounded above. By the completeness axiom it has a supremum, denoted by s. Then it is possible to show that $-s = \inf S$.

Problems

Problem 12.1.
Consider the sets below. For each one, decide whether the set is bounded above. If it is, give the supremum in $\mathbb{R}$. Then decide whether or not the set is bounded below. If it is, give the infimum. Finally, decide whether or not the supremum is a maximum, and whether or not the infimum is a minimum:
  (a) The closed interval $[0, 4]$;
  (b) The open interval $(0, 4)$;
  (c) The natural numbers $\mathbb{N}$;
  (d) The set $[0, \sqrt{2}] \cap \mathbb{Q}$;
  (e) The set $(0, \pi) \cap \mathbb{Q}$. (You may assume that $\pi \notin \mathbb{Q}$.)

Problem 12.2.
Consider the interval $(1, 4)$ in $\mathbb{R}$. Show in detail
  (a) that 4 is the supremum, and
  (b) that 1.1 is not a lower bound.

Problem 12.3.
Show that $\inf\{1/n : n \in \mathbb{Z}^+\} = 0$.

Problem 12.4.
Show that $\sup\{1 - 1/n : n \in \mathbb{Z}^+\} = 1$.

Problem 12.5.
Let S be a nonempty subset of $\mathbb{R}$, and $x \in \mathbb{R}$. Suppose that there exists $y \in S$ with $y < x$. Let $y_0 = \sup\{y \in S : y < x\}$.
  (a) Give an example of such a set S and real number x.
  (b) Show that $y_0 \leq x$.
  (c) Give an example to show that y_0 may equal x.
  (d) Give an example to show that y_0 may be strictly less than x.

Problem 12.6.
Let x and y be two real numbers. Prove that

$$\max\{x, y\} = \frac{|x - y| + x + y}{2}.$$

Problem° 12.7.
Let S be a nonempty subset of $\mathbb{R}$. Prove that S is bounded if and only if there exists $M \in \mathbb{R}$ such that $|x| \leq M$ for all $x \in S$.

Problem 12.8.
Let S be a nonempty bounded subset of $\mathbb{R}$. Show that $\inf S \leq \sup S$. Under what conditions on S would you have $\inf S = \sup S$?

Problem 12.9.
Let S be a nonempty bounded subset of $\mathbb{R}$ and let u be a real number such that $u < \sup S$. Show that there exists $s \in S$ such that $u < s$.

Problem 12.10.
Let S and T be nonempty bounded subsets of $\mathbb{R}$.
  (a) Show that $\sup(S \cup T) \geq \sup S$, and $\sup(S \cup T) \geq \sup T$.
  (b) Show that $\sup(S \cup T) = \max\{\sup S, \sup T\}$.
  (c) Try to state the results of (a) and (b) in English, without using mathematical symbols.

Problem 12.11.
Let $x \in \mathbb{R}$ and let S be a nonempty subset of $\mathbb{R}$ that is bounded above. We define a new set, $x + S$, by $x + S = \{x + s : s \in S\}$
  (a) Prove that $x + S$ is bounded above.
  (b) Prove that $x + \sup S$ is an upper bound of $x + S$. Conclude that $\sup(x + S) \leq x + \sup S$.
  (c) Prove that $x + \sup S = \sup(x + S)$.

Problem 12.12.
Let ϵ be a positive real number. Prove that for every real number a, there exists a rational number b (depending on a) such that $|a - b| < \epsilon$.

Problem 12.13.
Let $\sim$ denote a relation on a set S. The relation $\sim$ is called a **partial order** if the following three conditions are satisfied.
  (i) (Reflexive property) For all $x \in S$, we have $x \sim x$.
  (ii) (Transitive property) For all $x, y, z \in S$, if $x \sim y$ and $y \sim z$, then $x \sim z$.

(iii) (Antisymmetric property) For all $x, y \in S$, if $x \sim y$ and $y \sim x$, then $x = y$.

The relation $\sim$ is a **total order** on the set S if, in addition, (iv) below is satisfied.

(iv) For all $x, y \in S$, either $x \sim y$ or $y \sim x$.

(a) Show that the relation $x \sim y$ if and only if $x \leq y$ defines a total order on $\mathbb{R}$.

(b) Let A be a set containing at least two elements. We define an order on $\mathcal{P}(A)$ using the regular set inclusion $\subseteq$. Show that $(\mathcal{P}(A), \subseteq)$ is a partial order, but not a total order.

(c) Consider the relation $<$ on $\mathbb{R}$. Show that this is not a total order by exhibiting counterexamples for each total order property that is violated.

Problem♭ 12.14.

Prove that if a subset A of the reals has an infimum, then the infimum is unique.

Problem 12.15.

Prove that there exists $x \in \mathbb{R}^+$ with $x^2 = 3$.

Problem 12.16.

You showed in Problem 12.13 that $(\mathcal{P}(\mathbb{Z}), \subseteq)$ is a partial order. For every nonempty subset $\mathcal{A}$ of $\mathcal{P}(\mathbb{Z})$ we say that $U \in \mathcal{P}(\mathbb{Z})$ is an upper set of $\mathcal{A}$, if $X \subseteq U$ for all $X \in \mathcal{A}$. A nonempty set $\mathcal{A} \subseteq \mathcal{P}(\mathbb{Z})$ will be called an upper bounded set if there is an upper set of $\mathcal{A}$ in $\mathcal{P}(\mathbb{Z})$. We say $U_0 \in \mathcal{P}(\mathbb{Z})$ is a least upper set if (i) U_0 is an upper set of $\mathcal{A}$ and (ii) if U is another upper set of $\mathcal{A}$, then $U_0 \subseteq U$.

(a) Let $\mathcal{B} = \{\{1, 2, 5, 7\}, \{2, 8, 10\}, \{2, 5, 8\}\}$. Show that $\mathcal{B}$ is an upper bounded set and find a least upper set of $\mathcal{B}$, if there is one.

(b) Prove that every nonempty subset of $\mathcal{P}(\mathbb{Z})$ is upper bounded.

(c) Define "lower set," "lower bounded set," and "greatest lower set."

(d) Let $\mathcal{A}$ be a nonempty subset of $\mathcal{P}(\mathbb{Z})$. Using union and intersection, find an expression for least upper set of $\mathcal{A}$ and greatest lower set of $\mathcal{A}$.

(e) Prove that $(\mathcal{P}(\mathbb{Z}), \subseteq)$ has the "least upper set property" (in other words, show every upper bounded set has a least upper set).

Problem 12.17.
Prove that every bounded subset of $\mathbb{N}$ has a maximum.

Problem 12.18.
Prove that $(\mathbb{Z}, \leq)$ is complete in the following sense: If A is a nonempty set of integers that is bounded above, then there is an integer a such that $a = \sup A$.

Problem 12.19.
Prove the following statement: For every positive irrational real number a with $\sqrt{a} < 10000$, there is a positive integer n such that
$$\frac{10000}{n} < \sqrt{a} < \frac{10000}{n-1}.$$

Problem 12.20.
Suppose we define ∞ to be an object that satisfies $a \leq \infty$ for all $a \in \mathbb{R}$. Prove that $\infty \notin \mathbb{R}$.

Problem 12.21.
Let $a \in \mathbb{Q}$, $a \neq 0$ and $b \in \mathbb{R} \setminus \mathbb{Q}$. Prove the following:
  (a) $a + b \in \mathbb{R} \setminus \mathbb{Q}$;
  (b) $ab \in \mathbb{R} \setminus \mathbb{Q}$;
  (c) $1/b \in \mathbb{R} \setminus \mathbb{Q}$.

Problem 12.22.
Prove that if a is a rational number, then there is an irrational number b such that $a < b$.

Problem 12.23.
Prove that for two arbitrary real numbers a and b with $a < b$, there is an irrational number c such that $a < c < b$. (Hint: Consider $a/\sqrt{2}$ and $b/\sqrt{2}$.)

Tips: You Solved It. Now What?

Let's say we are now at the point where you solved the given problem and wrote up your first draft.

- Look over your solution. Does it use everything you are given?
- Is the answer reasonable?
- If there were places where you were unsure of your argument, check over those arguments carefully. You might find it helpful to write the solution, take a break, and then check the solution.
- Is your argument clear? Did you choose your notation well? Did you introduce all notation before you used it?
- Is there a shorter or more intuitive argument?
- Do you fully understand what you did? Spend some time thinking about the method and what you proved. Could you have gotten a better result?
- When do these methods work? What are the restrictions? Where have you seen them before?
- If the problem was hard for you to solve, what made it hard? What were the important ideas that you were missing?
- This is a good opportunity to learn about yourself, too. Which problems do you like best? Why?

13

CHAPTER

Functions, Domain, and Range

What is a function? You've probably gotten a definition of this somewhere along the way. We will state the definition of function in terms of relations, which is probably different than the way you have seen it stated.

Let A and B be sets. A **function** f from A to B is a relation from A to B satisfying

(*i*) for all $a \in A$, there exists $b \in B$ such that $(a, b) \in f$, and
(*ii*) for all $a \in A$, and all $b, c \in B$, if $(a, b) \in f$ and $(a, c) \in f$, then $b = c$.

A function is often called a **map** or **mapping**. We usually write $f : A \to B$ to indicate that f is a function from A to B. The two conditions above define a function. When they are satisfied, we say the function is **well-defined**. If the object we try to define does not satisfy these properties, it isn't a function, and we often say that f (which we shouldn't call a function) is not well-defined.

Condition (*i*) makes sure that each element in A is related to some element of B, while condition (*ii*) makes sure that no element of A is related to more than one element of B. Note that it may be the case that an element of B has no element of A that it is related to; or an element of B could be related to more than one element of

147

A. The set A is called the **domain**, and denoted by dom(f), and the set B is called the **codomain**, and denoted cod(f).

As a first example, consider the function that assigns to a citizen of the United States his or her height measured on a particular day at a particular time. This is a function because (i) each person has a height, and (ii) each person has exactly one height on that day, at that time. Now let's turn to a nonexample. We still consider the domain to be the set of citizens of the United States, but this time let the codomain consist of all the countries in the world (on a particular day, at a particular time). Consider the relation that assigns to each person in the domain his or her country (countries) of citizenship. This is not a function because a United States citizen can be a citizen of more than one country. Though (i) is satisfied because each person in the domain is a U.S. citizen, (ii) is not.

Exercise 13.1.
Let $A = \{1, 2, 3\}$ and $B = \{2, 4, 6\}$. Which of the following are functions from A to B? If they are not functions, explain which rule is violated.
  (a) The relation f is $\{(1, 2), (2, 4), (3, 4)\}$.
  (b) The relation f is $\{(1, 2), (1, 4), (2, 2), (3, 6)\}$.
  (c) The relation f is $\{(1, 2), (3, 4)\}$.
  (d) The relation f is $\{(2, 4), (1, 2), (3, 6)\}$. ○

Exercise 13.2.
You probably learned that a function $f : \mathbb{R} \to \mathbb{R}$ can be represented by a graph, and that there is a vertical line test to determine whether or not f is a function (see Figure 13.1). Which condition in the definition corresponds to the vertical line test? Why? ○

In Exercise 13.1, you probably recognized (d) as a function. It is more usual to write $f(1) = 2$, $f(2) = 4$, and $f(3) = 6$. Since each x in the domain is related to a unique y in the codomain, we will write $f(x) = y$ rather than $(x, y) \in f$.

Here are some more examples and nonexamples of functions.

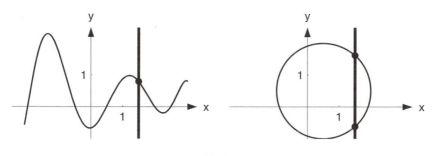

FIGURE 13.1

Exercise 13.3.
Decide which of the following are functions and which are not, giving reasons for your answers.
(a) Let $f : \mathbb{R} \to \mathbb{R}$ be defined by $f(x) = 3x + 2$.
(b) Let $f : \mathbb{R} \to \mathbb{R}$ be defined by $f(x) = 1/x^2$.
(c) Let $f : \mathbb{R} \to \mathbb{R}^2$ be defined by $f(x) = (x, x)$.
(d) Let $f : \mathbb{Q} \to \mathbb{Q}$ be defined by $f(p/q) = 1/q$, where p and q are integers and $q \neq 0$.
(e) Let $f : \mathbb{R}^2 \to \mathbb{R}^2$ be defined by $f(x, y) = (x, 3)$. $\bigcirc$

You have seen many examples of functions. One particular type of example, that of a function defined in cases, allows us to explicitly illustrate many of the ideas discussed in this section. Before you begin working with a function that is defined in cases, make sure that you understand the function. If you can, graph it. Remember that the best thing to do is to work with concrete objects (like trying $x = 2$ or $x = -3$) until you get a feel for what is happening. For functions that are defined in cases we have to be particularly careful to check that the cases don't overlap; or if they do, that the function is defined in a unique way for all the elements in the domain that are in the overlap. Of course, we are not changing the rules here. All you really have to do is check that conditions (i) and (ii) of the definition hold. Here are some examples.

Example 13.4.
We will check to see whether each of the objects defined below and graphed in Figure 13.2 is a function.

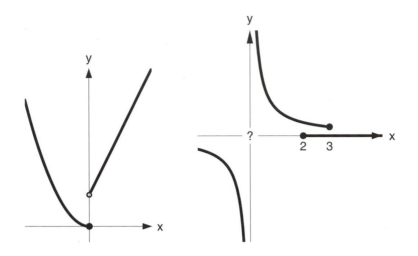

FIGURE 13.2 The graph on the left is of f, the one on the right is of g.

(a) Let $f : \mathbb{R} \to \mathbb{R}$ be defined by

$$f(x) = \begin{cases} x^2 & \text{if } x \leq 0 \\ 2x + 1 & \text{if } x > 0 \end{cases}.$$

(b) Let $g : \mathbb{R} \to \mathbb{R}$ be defined by

$$g(x) = \begin{cases} 0 & \text{if } x \geq 2 \\ 1/x & \text{if } x \leq 3 \end{cases}.$$

For (a) note that the domain is $\mathbb{R}$ and the codomain is also $\mathbb{R}$. From the definition of f it is easy to see that f is defined for all $x \in \mathbb{R}$. Hence condition (*i*) of the definition of a function holds.

Now let $a \in \mathbb{R}$ and suppose that there exist real numbers b and c with $f(a) = b$ and $f(a) = c$. The most orderly way to check condition (*ii*) is the following: If $a \geq 0$, then $b = f(a) = a^2$ and $c = f(a) = a^2$, so $b = c$. If $a < 0$, then $b = f(a) = 2a + 1$ and $c = f(a) = 2a + 1$. Hence $b = c$. In either case, $b = c$. So condition (*ii*) holds. Since both (*i*) and (*ii*) are satisfied, f is well-defined.

The formula given in part (b) does not define a function for two reasons. First note that 0 is in the domain. Since $0 \leq 3$, we see that $g(0)$ is not defined to be an element in the codomain. Hence condition (*i*) is violated, and we conclude that g is not a function.

We mention here that there is a second problem with the definition of g: consider a real number a such that $2 \leq a \leq 3$. For instance, let $a = 2.5$. Then $g(2.5) = 0$ (since $2.5 \geq 2$) and $g(2.5) = 2/5$ (since $2.5 \leq 3$). This violates condition (ii) of the definition of a function. Thus g does not satisfy condition (i) or (ii). Therefore, the object defined above is not a function, for two reasons. ○

Although the example above violates both (i) and (ii), keep in mind that it is enough that (i) or (ii) alone be violated to assure that f is not a function.

Exercise 13.5.
For each of the two examples below decide whether or not the object so defined is a function. Give reasons for your answers.
  (a) Let $f : \mathbb{R} \to \mathbb{R}$ be defined by

$$f(x) = \begin{cases} x^2 & \text{if } x \geq 0 \\ -(x^2) & \text{if } x \leq 0 \end{cases}.$$

  (b) Let $f : \mathbb{Z} \to \mathbb{Z}$ be defined by

$$f(x) = \begin{cases} 1 & \text{if } x \in 2\mathbb{Z} \\ 2 & \text{if } x \text{ is prime} \\ 3 & \text{otherwise} \end{cases}.$$

○

One very important example of a function defined in cases is the familiar absolute value function.

Example 13.6.
The absolute value function $f : \mathbb{R} \to \mathbb{R}$ is defined by $f(x) = |x|$. It is easy to check that this does define a function on $\mathbb{R}$.

When you define a new mathematical concept, it's always a good idea to think about it and pose questions. Of course, it's also a good idea to answer those questions, if you can. We now turn to some questions that we find interesting. See if you can think of some questions on your own.

What does it mean to say that two functions f and g are equal? Since this is a very important concept that we will need again later, we provide the answer here. But try and think about how this answer follows from the definition of a function.

> Two functions f and g are equal if $\text{dom}(f) = \text{dom}(g)$ and $f(x) = g(x)$ for all $x \in \text{dom}(f)$.

Here's a second question: What is the function's relationship to elements of the domain, and how does this differ from the function's relationship to elements of the codomain? We must be able to evaluate f for every element in the domain, while elements in $\text{cod}(f)$ may or may not be associated with elements of the domain. The elements of the codomain that are related to elements of $\text{dom}(f)$ are obviously important in understanding the function. For this reason, we look at the set called the range of f, which consists precisely of these points.

Given a function $f : A \to B$, the **range** of f, denoted $\text{ran}(f)$, is defined by

$$\text{ran}(f) = \{b \in B : \text{there exists at least one } a \in A \text{ such that } f(a) = b\}.$$

Sometimes it is fairly easy to determine the range, but it generally requires a method (demonstrated below) that we think of as working backwards. You'll start with $b \in B$ and try to find $a \in A$. Then, to show that $b \in \text{ran}(f)$, you have two things to check: The element a must map to b under f (that is, $f(a) = b$), *and* a must be an element of A. This latter statement is often obvious, but don't forget to check it!

It's always easier to start with small sets and see if you understand what is happening. You can do this visually as well. For example, say $A = \{1, 2, 3\}$, $B = \{2, 3, 4\}$, and the function $f : A \to B$ is defined by $f(1) = 3, f(2) = 3$ and $f(3) = 4$. We can "see" the action of f by drawing a little picture as in Figure 13.3

From this picture we can see easily that f sends two things to 3, one thing to 4, and nothing to 2. So we can "see" that though 2 is in the codomain, it is not in the range. If you are asked for examples or counterexamples, remember that small sets will sometimes do the trick!

Our next example is really a method. Once we complete the example, we will review exactly what you must do in similar cir-

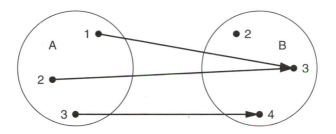

FIGURE 13.3

cumstances. But one thing we will mention in advance: you will always need to devise a plan as we do below.

Example 13.7.
Let $f : \mathbb{R} \setminus \{1\} \to \mathbb{R}$ be defined by $f(x) = (x+1)/(x-1)$. Determine the range of f.

  "*Devising a plan.*" We need to figure out which $y \in \mathbb{R}$ come from something under f. It's a bit difficult to simply gaze at f, or even the graph of f, and see what comes out of it, so we'll try working backwards to see what y might be. (Though the graph of f provides a good way to see if your answer is reasonable, it does not provide a proof.) So, suppose $y \in \mathbb{R}$ did come from something in the domain. That would mean

$$y = f(x), \text{ for some } x \in \mathbb{R} \setminus \{1\};$$

in other words, $y = (x+1)/(x-1)$. Since we need to figure out what x is, we should solve for it. Multiplying through by $x - 1$, we get $(x + 1) = yx - y$. Collecting all terms involving x yields $x - yx = -y - 1$. Factoring out x, dividing, simplifying, and ignoring potential problems (like what?), we get $x = (y+1)/(y-1)$. So if y came from some x at all, y had to come from $x = (y+1)/(y-1)$. That's fine, as long as $y \neq 1$ (that was a potential problem). So ran(f) "appears to be" $\{y \in \mathbb{R} : y \neq 1\} = \mathbb{R} \setminus \{1\}$.
  The reason for saying "appears to be" is that we started by assuming y came from something called x, and then found out what x had to be. But the definition of range really requires us to start with an x and show that $f(x) = y$. So we need to check that everything we did

above is reversible, and that the two sets ran(f) and $\mathbb{R} \setminus \{1\}$ are equal. All of this was helpful in deciding what the range is, but the actual proof is still to come. The proof below is the form you should follow. When we write it, we need to pretend the reader has not seen the work we just completed.

Proof.
We will show that ran(f) $= \mathbb{R} \setminus \{1\}$. Let $y \in$ ran(f). Then, clearly, $y \in \mathbb{R}$. So ran(f) $\subseteq \mathbb{R}$. To show that $y \neq 1$, suppose that this is not the case; so we will suppose $y = 1 \in$ ran(f) and see what happens. Since $y \in$ ran(f), there exists a point x in the domain with $f(x) = y = 1$. Using the definition of f, we find that $1 = f(x) = (x+1)/(x-1)$. Therefore, $x + 1 = x - 1$. This would mean that $1 = -1$, which is not possible. So $y \in$ ran(f) implies $y \in \mathbb{R}$ and $y \neq 1$. Thus, ran(f) $\subseteq \mathbb{R} \setminus \{1\}$.

Now let $y \in \mathbb{R} \setminus \{1\}$. Let $x = (y+1)/(y-1)$. Since $y \neq 1$, we see that $x \in \mathbb{R}$. Remember that we need to check that $x \in$ dom(f). We know that $x \in \mathbb{R}$. Could we possibly have $x = 1$? Suppose we do, then $1 = (y+1)/(y-1)$ which implies $y - 1 = y + 1$. Thus we would have $-1 = 1$, which is impossible. So $x \in$ dom(f) and we can evaluate f at x to obtain

$$f(x) = \frac{\frac{y+1}{y-1} + 1}{\frac{y+1}{y-1} - 1} = \frac{y+1+y-1}{y+1-y+1} = y.$$

It follows that $\mathbb{R}\setminus\{1\} \subseteq$ ran(f). Therefore ran(f) $= \mathbb{R}\setminus\{1\}$, completing the proof. ∎

Before going on, we will make two remarks. If you hadn't read *"Devising a plan"* above the proof, the definition of $x = (y+1)/(y-1)$ would probably look bizarre. Remember that we didn't guess it; we worked backwards to see what x had to be. One other thing to note is that ran(f) $\neq \mathbb{R}$, but ran(f) $= \mathbb{R} \setminus \{1\}$. So f maps into $\mathbb{R}$ but it doesn't "hit" the value 1. We'll come back to this in the next chapter. ○

So what must we do when we have to find the range of a function? First, we need to take out a different sheet of paper and figure out what the set should be. Let's say we decide the range is a set called B. Then we need to show the reader that the two sets are equal. There are often many ways to do it, but one way is to start with an element in the range (tell the reader you are doing this) and show it is in B. Then start with an element y in B (tell the reader you are doing this, too) and find an x (which you found somewhere else, but the reader doesn't necessarily need to see that) that satisfies two things: x is in the domain of your function and $f(x) = y$. Write your proof up carefully, identifying variables before you use them, and always checking that your variables are in the appropriate sets.

Exercise 13.8.
What is the range of each of the functions below? A picture, when appropriate, is a lovely addition and is heartily encouraged. It does not, however, substitute for the real thing. Write out everything explicitly.
  (a) The function $f : \mathbb{R} \setminus \{0\} \rightarrow \mathbb{R}$ defined by $f(x) = 1/x$.
  (b) The function $f : \mathbb{Z} \times (\mathbb{Z} \setminus \{0\}) \rightarrow \mathbb{R}$ defined by $f(x, y) = x/y$.
  (c) The function $f : \mathbb{R} \rightarrow \mathbb{R}$ defined by $f(x) = x^2 + 4x + 5$. ○

Solutions to Exercises

Solution to Exercise (13.1).
The relations in (a) and (d) are functions, those in (b) and (c) are not.

Solution to Exercise (13.2).
Condition (ii) corresponds to the vertical line test, since it says that if we draw the vertical line $x = a$, it should pass through the graph of f at most once.

Solution to Exercise (13.3).
Parts (a), (c), and (e) define functions. The others do not. In (b), we have not defined $f(0)$ as an element of $\mathbb{R}$. In (d) note that if, for example, we consider $a = 2/1 = 4/2$, then $f(2/1) = 1$, while $f(4/2) = 1/2$. Thus $(2, 1)$ and $(2, 1/2)$ are both elements of the relation and condition (ii) is violated.

Solution to Exercise (13.5).
For part (a), both conditions in the definition of function are satisfied. Note that though $x = 0$ appears twice in the definition of f, in both cases $f(0) = 0$. For part (b), consider $x = 2$. Since $x = 2 \in 2\mathbb{Z}$, we have $f(2) = 1$. On the other hand, 2 is also prime, so $f(2) = 2$. Thus $(2, 1)$ and $(2, 2)$ are both elements of the relation, but $1 \neq 2$, and condition (ii) is violated.

Solution to Exercise (13.8).
For (a) we claim that $\text{ran}(f) = \mathbb{R} \setminus \{0\}$. It is clear that $\text{ran}(f) \subseteq \mathbb{R} \setminus \{0\}$. So suppose that $y \in \mathbb{R} \setminus \{0\}$. Let $x = 1/y$. Then $x \in \mathbb{R}$ and $x \neq 0$. Thus, $x \in \text{dom}(f)$. Furthermore, $f(x) = 1/(1/y) = y$. Therefore, $y \in \text{ran}(f)$ and $\mathbb{R} \setminus \{0\} \subseteq \text{ran}(f)$, completing the proof.

For (b), you should show that $\text{ran}(f) = \mathbb{Q}$.

For (c) we claim that $\text{ran}(f) = \{z \in \mathbb{R} : z \geq 1\}$. (We went to another sheet of paper to come up with this claim. A sketch (see Figure 13.4) is also helpful here, but it is *not* a proof.)

Proof.
First note that if $y \in \text{ran}(f)$, then there exists $x \in \mathbb{R}$ such that $y = x^2 + 4x + 5$. Completing the square, we get $y = (x + 2)^2 + 1$. Since $(x + 2)^2 \geq 0$, we see that $y \geq 1$. Therefore, $y \in \{z \in \mathbb{R} : z \geq 1\}$, and hence $\text{ran}(f) \subseteq \{z \in \mathbb{R} : z \geq 1\}$. Now suppose that $y \in \{z \in \mathbb{R} : z \geq 1\}$. Let $x = \sqrt{y - 1} - 2$. (We worked backwards to get this, of course.) Since $y \geq 1$, we have $x \in \mathbb{R}$. So $x \in \text{dom}(f)$. Furthermore, $f(x) = f(\sqrt{y-1}-2) = (\sqrt{y-1}-2)^2 + 4(\sqrt{y-1}-2) + 5$. Thus (as the

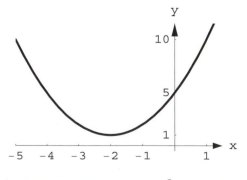

FIGURE 13.4 $f(x) = x^2 + 4x + 5$

reader can check) $f(x) = y - 1 - 4\sqrt{y-1} + 4 + 4\sqrt{y-1} - 8 + 5 = y$.
Therefore, $y \in \text{ran}(f)$ and $\{z \in \mathbb{R} : z \geq 1\} \subseteq \text{ran}(f)$, as desired. ∎

Spotlight: The Definition of Function

It's probably difficult to imagine that there could be any debate about
the definition of function. In fact, the development of the definition
of function is quite interesting. For example, Leonhard Euler first
defined a function as follows [73, p. 72] "A function of a variable
quantity is an analytical expression composed in any manner from
that variable quantity and numbers or constant quantities." Euler
later revised his definition because of work on a problem known
as the vibrating string problem. Discussion ensued, and Dirichlet is
now often credited with providing us with roughly the definition we
use today.

Once this discussion appeared to be settled, people could then
concentrate on studying various kinds of functions; including, for example, continuous, discontinuous, differentiable, or even nowhere
differentiable functions. Dirichlet also introduced the following
example (now called the Dirichlet function):

$$D(x) = \begin{cases} c & \text{if } x \in \mathbb{Q} \\ d & \text{if } x \in \mathbb{R} \setminus \mathbb{Q} \end{cases},$$

where c and d are distinct real numbers. This was the first example of many things, including the first example of a function that
is discontinuous everywhere (see [47]). In a very interesting article
written around 1940 (or, rather, the English translation of this article), Luzin points out that not everyone agreed that Dirichlet had
completely answered the question of what a function is. According
to Luzin [53, p. 263], some mathematicians found the definition perfect, others found it too broad, and still others found it meaningless.
Even as late as 1928, Hermann Weyl [84, p. 22] stated that no one
can explain what a function is; then Weyl finishes the paragraph by
telling us what a function is: "A function is given if by some definite
rule to each real number a there is assigned a real number b (as e.g.

by the formula $b = 2a + 1$). One then says that b is the value of the function f for the value a of the argument."[1]

For an overview of the definition of the concept of function, we recommend Rüthing's entertaining paper [73], where definitions (from 1718 to 1939) attributed to various authors are presented in their original language, with translation and without comment. You will notice that the final definition, due to N. Bourbaki and given in 1939, agrees with our definition.

The history of the vibrating string problem is described in [48, pp. 503–518]. In [46, p. 724], Katz presents the definition of function used by Johann Bernoulli, an earlier and later definition used by Euler, and definitions attributed to Lacroix, Fourier, Heine, and Dedekind. For a complete and readable overview on this topic, we recommend the papers of Luzin (both [52] and [53]), Youschkevitch [87], and Kleiner [47]. Kleiner's paper also has an extensive bibliography.

Problems

Problem 13.1.
Complete the following: A relation $f : A \to B$ is not a function if . . .

Problem 13.2.
Suppose that $f : X \to Y$. Recall that the definition of ran(f) was stated in the text. State carefully what it means when we say $y \in Y$ is not in the range of f.

Problem 13.3.
Which of the following are functions? Give reasons for your answers.
 (a) Define f on $\mathbb{R}$ by $f = \{(x, y) : x^2 + y^2 = 4\}$.
 (b) Define $f : \mathbb{R} \to \mathbb{R}$ by $f(x) = 1/(x + 1)$.
 (c) Define $f : \mathbb{R}^2 \to \mathbb{R}$ by $f(x, y) = x + y$.
 (d) The domain of f is the set of all closed intervals of real numbers of the form $[a, b]$, where $a, b \in \mathbb{R}$, $a \le b$, and f is defined by $f([a, b]) = a$.

---

[1] The translation is ours.

(e) Define $f : \mathbb{N} \times \mathbb{N} \to \mathbb{R}$ by $f(n, m) = m$.

(f) Define $f : \mathbb{R} \to \mathbb{R}$ by

$$f(x) = \begin{cases} 0 & \text{if } x \geq 0 \\ x & \text{if } x \leq 0 \end{cases}.$$

(g) Define $f : \mathbb{Q} \to \mathbb{R}$ by

$$f(x) = \begin{cases} x + 1 & \text{if } x \in 2\mathbb{Z} \\ x - 1 & \text{if } x \in 3\mathbb{Z} \\ 2 & \text{otherwise} \end{cases}.$$

(h) The domain of f is the set of all circles in the plane $\mathbb{R}^2$ and, if c is such a circle, define f by $f(c) = $ the circumference of c.

(i) *(For students with a background in calculus.)* The domain of f is the set of all polynomials with real coefficients, and f is defined by $f(p) = p'$. (Here p' is the derivative of p.)

(j) *(For students with a background in calculus.)* The domain of f is the set of all polynomials and f is defined by $f(p) = \int_0^1 p(x)\, dx$. (Here $\int_0^1 p(x)\, dx$ is the definite integral of p.)

Problem 13.4.

Let $f : \mathcal{P}(\mathbb{R}) \to \mathbb{Z}$ be defined by

$$f(A) = \begin{cases} \min(A \cap \mathbb{N}) & \text{if } A \cap \mathbb{N} \neq \emptyset \\ -1 & \text{if } A \cap \mathbb{N} = \emptyset \end{cases}.$$

Prove that f above is a well-defined function.

Problem 13.5.

Let X be a nonempty set and let A be a subset of X. We define the **characteristic function** of the set A by

$$\chi_A(x) = \begin{cases} 1 & \text{if } x \in A \\ 0 & \text{if } x \in X \setminus A \end{cases}.$$

(a) Since this is called the characteristic function, it probably is a function, but check this carefully anyway.

(b) Determine the domain and range of this function. Make sure you look at all possibilities for A and X.

Problem 13.6.
Let X be a bounded nonempty subset of $\mathbb{R}$. If we define $g : \mathcal{P}(X) \setminus \{\emptyset\} \to \mathbb{R}$ by $g(S) = \sup S$, is g a well-defined function? Why or why not?

Problem 13.7.
Consider the (well-defined) function $f : \mathbb{R} \setminus \{3/2\} \to \mathbb{R}$ defined by $f(x) = (x - 5)/(2x - 3)$. Carefully prove that $\text{ran}(f) = \mathbb{R} \setminus \{1/2\}$.

Problem 13.8.
(a) Give an example of a function f from $\mathbb{R} \times \mathbb{R}$ to $\mathbb{R}^+$.
(b) Give an example of a function f from $\mathbb{Z} \times \mathbb{Z}$ to $\mathbb{N}$ such that $\text{ran}(f) = \mathbb{N}$. (Prove that f is a function and $\text{ran}(f) = \mathbb{N}$.)
(c) Give an example of a function f from $\mathbb{Z} \times \mathbb{Z} \to \mathbb{N}$ such that $\text{ran}(f) \neq \mathbb{N}$. (Prove that f is a function and $\text{ran}(f) \neq \mathbb{N}$.)

Problem 13.9.
Let a, b, c, and d be real numbers with $a < b$ and $c < d$. Let $[a, b]$ and $[c, d]$ be two closed intervals. Find a function f such that $f : [a, b] \to [c, d]$ and $\text{ran}(f) = [c, d]$. Prove everything.

Problem 13.10.
(a) Define $f : \mathbb{Z} \to \mathbb{N}$ by $f(x) = |x|$. Is f a function? If so, determine $\text{ran}(f)$.
(b) Define $f : \mathbb{R}^2 \to \mathbb{R}$ by $f(x, y) = x$. Is f a function? If so, determine $\text{ran}(f)$.

Problem 13.11.
Suppose that f is a function from a set A to a set B. Thus, we know that f is a subset of $A \times B$. Is the relation $\{(y, x) : (x, y) \in f\}$ necessarily a function from B to A? Why or why not? (Say as much as is possible to say with the given information.)

Problem 13.12.
Which of the following functions equal $f : \mathbb{Z} \to \mathbb{Z}$ defined by $f(x) = |x|$? Prove your answers (make sure you show that the functions below either equal f or do not equal f).
(a) The function $g : \mathbb{Z} \to \mathbb{R}$ defined by $g(x) = |x|$.
(b) The function $h : \mathbb{Z} \to \mathbb{Z}$ defined by $h(x) = \sqrt{x^2}$.

(c) The function $k : \mathbb{R} \to \mathbb{R}$ defined by $k(x) = \sqrt{x^2}$.

(d) The function $l : \mathbb{N} \to \mathbb{Z}$ defined by $l(x) = \sqrt{x^2}$.

Problem 13.13.

Let X be a nonempty set. Find all relations on X that are both equivalence relations and functions.

Problem 13.14.

We can now define an indexed family of sets more rigorously than we did in Chapter 8. Let I and X be sets and $f : I \to \mathcal{P}(X)$ be a function. Then ran(f) is called an **indexed family** of subsets of X.

As a specific example, let $f : \mathbb{Z}^+ \to \mathcal{P}(\mathbb{R})$ be defined by $f(n) = \{x \in \mathbb{R} : \pi - 2n \le x \le \pi + 2/n\}$.

(a) Find $\bigcup_{n \in \mathbb{Z}^+} f(n)$.

(b) Find $\bigcap_{n \in \mathbb{Z}^+} f(n)$.

14

CHAPTER

Functions, One-to-One, and Onto

Functions can map elements from the domain to the codomain in many ways. A function may "hit" every element in the codomain, or it may "miss" some. It may assign more than one x to a y or it may assign exactly one x to each y. We will understand our function better if we know which of these things it does. Precise formulations of these ideas will be given in a moment. It's a mouthful, though, and really requires practice.

To say that a function $f : A \rightarrow B$ is **one-to-one** means that for all $a_1, a_2 \in A$, if $f(a_1) = f(a_2)$, then $a_1 = a_2$. A function $f : A \rightarrow B$ is **onto** if $\mathrm{ran}(f) = B$. If a function has this property, then we say that f maps A onto B. Some authors use the word **injective** rather than one-to-one and **surjective** rather than onto. If a function is both one-to-one and onto (or injective and surjective), then we say the function is **bijective**.

Diagrams using small sets may help illustrate the ideas involved in these definitions. We sketch two functions that are not one-to-one in Figure 14.1 and two functions that are not onto in Figure 14.2. Can you make a diagram for a function that is bijective?

Before moving on, let's think about these definitions. The definition of one-to-one is an implication with quantifiers on elements of the domain A. It moves "forward," in the sense that we start with

163

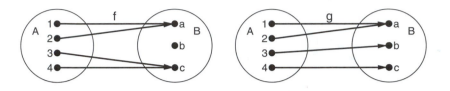

FIGURE 14.1   The function f does not map onto B, but the function g does. Neither function is one-to-one.

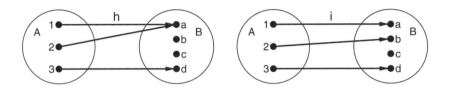

FIGURE 14.2   The function h is not one-to-one, but the function i is. Neither function maps onto B.

elements in A and see where f maps them. The definition of onto, on the other hand, requires that we show something about every element in the set B. It will require us to move "backward," in the sense that we will start with something in B and see what element of A is mapped to it under f.

If you want to show that a function is onto, we said that you must check that ran$(f) = B$. Technically that would mean showing ran$(f) \subseteq B$ and $B \subseteq$ ran(f). But if we are showing that a *function* $f : A \to B$ is onto, we already know that ran$(f) \subseteq B$; that is, we get that half for free. So if we know we have a function $f : A \to B$ (maybe it was given to us; maybe we showed it), to show that f maps A onto B, we only have to check that $B \subseteq$ ran(f). And what does this mean? The answer is given in the lemma below.

Lemma 14.1.
Let $f : A \to B$ be a function. Then f maps A onto B if and only if for all $b \in B$ there exists (at least one) $a \in A$ such that $f(a) = b$.

Since this lemma is just a reformulation of the definition of onto, we will often use it without explicitly referencing it.

Let's run through some of these ideas in slow motion. We'll begin with a simple example, and move on to a more challenging one.

Example 14.2.
Let $f : \mathbb{R} \to \mathbb{R}$ be defined by $f(x) = 2x + 5$. Then f is one-to-one.

Proving a function is one-to-one is often easier than proving it is onto, because one-to-one doesn't have that backwards quality. To prove it, we assume that (for two arbitrary points x_1 and x_2 in the domain) $f(x_1) = f(x_2)$, and show that $x_1 = x_2$. So now we'll just dive in here. Don't forget that the definition of one-to-one is an implication and therefore we expect to use our assumption.

Proof.
Let $x_1, x_2 \in \mathbb{R}$. If $f(x_1) = f(x_2)$, then $2x_1 + 5 = 2x_2 + 5$. Simplification yields $x_1 = x_2$, as desired. Therefore f is one-to-one. ∎

We turn to a more interesting example. Note that even though the functions in these two examples are quite different, the proofs that they are one-to-one are quite similar.

Example 14.3.
In Example 13.7, we considered $f : \mathbb{R} \setminus \{1\} \to \mathbb{R}$ by $f(x) = (x + 1)/(x - 1)$. We started with an element y and solved for x. We found an x that corresponded to y, but there may be others. Is the function one-to-one (meaning there aren't any others)?

We claim that $f : \mathbb{R} \setminus \{1\} \to \mathbb{R}$ defined by $f(x) = (x + 1)/(x - 1)$ is one-to-one.

Proof.
Let $x_1, x_2 \in \mathbb{R} \setminus \{1\}$. If $f(x_1) = f(x_2)$, then

$$\frac{x_1 + 1}{x_1 - 1} = \frac{x_2 + 1}{x_2 - 1}.$$

Multiplying through yields $x_1 x_2 + x_2 - x_1 - 1 = x_2 x_1 + x_1 - x_2 - 1$. Cancelling, we find that $x_1 = x_2$, as desired. Therefore f is one-to-one. ∎

Exercise 14.4.
What do you need to do in order to show that a function is not one-to-one? Use what you just decided to show that the function $f : \mathbb{R} \to \mathbb{R}$ defined by $f(x) = x^2$ is not one-to-one.   ○

Exercise 14.5.
In Exercise 13.8, which of the functions are one-to-one? If they are not one-to-one, show that carefully as well.   ○

It is now time to investigate what it really means when we say that a function maps a set A onto a set B.

Example 14.6.
Prove that the function $f : \mathbb{R} \to \mathbb{R}$ defined in Example 14.2 by $f(x) = 2x + 5$ is onto.

We first devise our plan.

"Devising a plan." We have checked that $f : \mathbb{R} \to \mathbb{R}$ is well-defined, so (in view of Lemma 14.1) we let $y \in \mathbb{R}$. We must show that $y = f(x)$ for some $x \in \mathbb{R}$. Thus, we must show that $y = 2x + 5$ for some $x \in \mathbb{R}$. It is now easy to see that $x = (y - 5)/2$ will work. Remember, when we write this up, we will act as though the reader has not seen this work.

Proof.
Let $y \in \mathbb{R}$ and let $x = (y - 5)/2$. Then $x \in \mathbb{R}$ and

$$f(x) = 2 \left(\frac{y - 5}{2} \right) + 5 = y.$$

Therefore, $y \in \mathrm{ran}(f)$ and we have shown that $\mathbb{R} \subseteq \mathrm{ran}(f)$. Since $f : \mathbb{R} \to \mathbb{R}$ is a well-defined function, f maps $\mathbb{R}$ onto $\mathbb{R}$.   ■

Functions that are defined in cases will play an important role in the rest of this course. They are also illuminating examples, because they show just how much one-to-one depends on "what goes into f" and just how much onto depends on "what comes out." Showing

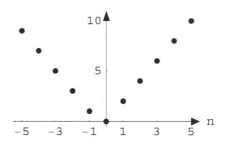

FIGURE 14.3

that they are one-to-one and onto is a bit tricky. We'll go through one example carefully first.

Example 14.7.
We will show that $f : \mathbb{Z} \to \mathbb{N}$ as defined below and graphed in Figure 14.3 is a bijective function:

$$f(n) = \begin{cases} 2n & \text{if } n \geq 0 \\ -2n - 1 & \text{if } n < 0 \end{cases}.$$

We should mention something before we begin. We, the authors of this text, checked that the f in this example is really a function from $\mathbb{Z}$ to $\mathbb{N}$. But don't trust us; check it again. Remember to check that f assigns a value in $\mathbb{N}$ to each element of $\mathbb{Z}$, and that f assigns at most one value in $\mathbb{N}$ to each integer. We now begin our example.

"*Devising a plan.*" Now to prove that f is one-to-one, we must show that if m and n are integers, and $f(m) = f(n)$, then $m = n$. But it's sort of confusing here, since we have more than one choice for $f(m)$ and $f(n)$. The problem seems to be that the function is defined in cases, and it's not immediately clear which case to use. The choice depends on whether m and n are negative or nonnegative. So we'll break this into all possible cases. Let's see. Both could be nonnegative (case 1), both negative (case 2), or one nonnegative, one negative (case 3). So let's check them all.

Proof that f is one-to-one.
Let $m, n \in \mathbb{Z}$ and suppose that $f(m) = f(n)$.

Case 1. Suppose that $m \geq 0$ and $n \geq 0$. Then $f(m) = 2m$ and $f(n) = 2n$. Thus $2m = 2n$, and therefore $m = n$.

Case 2. Suppose that $m < 0$ and $n < 0$. Then $f(m) = -2m - 1$ and $f(n) = -2n - 1$. Thus, $-2m - 1 = -2n - 1$, and therefore $m = n$.

Case 3. Suppose that one of the two, say m, is nonnegative, and the other is negative. Then $f(m) = 2m$ and $f(n) = -2n - 1$. Thus $2m = -2n - 1$. But this means that an even number, $2m$, is equal to an odd number, $-2n-1$, which is impossible.

Therefore, if $f(m) = f(n)$, only case 1 and case 2 can occur. In either of these cases, we have shown that $m = n$. Thus f is one-to-one. ∎

Now we will show that f maps $\mathbb{Z}$ onto $\mathbb{N}$.

"Devising a plan." Recall that to show f is onto, we must show that $\operatorname{ran}(f) = \mathbb{N}$. We (you, actually) already checked that f is a well-defined function from $\mathbb{Z}$ to $\mathbb{N}$, so $\operatorname{ran}(f) \subseteq \mathbb{N}$. So let $k \in \mathbb{N}$. We wish to find m such that $f(m) = k$ and $m \in \mathbb{Z}$. Again, because there are cases, this is a bit confusing. So let's try to think about this before we really begin. We'll try to do this for a few specific points. Though this won't prove anything, it may tell us how to begin our proof. So moving along in the spirit of showing something is onto, if $f(m) = 4$, what's m? (It's 2.) If $f(m) = 3$, what's m? (It's −2.) Now maybe you see it. If you don't, keep trying until you do. After a while, you should see that our choice of m depends on whether k is even or odd.

Proof that f maps $\mathbb{Z}$ onto $\mathbb{N}$.

Let $k \in \mathbb{N}$. If k is even, then $k = 2m$ for some $m \in \mathbb{Z}$ with $m \geq 0$. Thus, $m \in \mathbb{Z}$ and $f(m) = 2m = k$. If k is odd, then $k + 1$ is even. Hence $m = (k + 1)/(-2) \in \mathbb{Z}$. Since $k \geq 1$, we have $m < 0$. Thus, $f(m) = -2m - 1 = -2((k + 1)/(-2)) - 1 = k$. We conclude that for all $k \in \mathbb{N}$, there exists $m \in \mathbb{Z}$ such that $f(m) = k$. Since $f : \mathbb{Z} \to \mathbb{N}$ is a well-defined function, f maps $\mathbb{Z}$ onto $\mathbb{N}$. ∎

Note that though the functions in the last two examples are quite different, the proofs that they are onto are quite similar.

What about an example of a function $f : A \to B$ that does not map A onto B? Our Lemma 14.1 is quite handy here: Let $f : A \to B$ be a function. Then f is not onto if there exists $b \in B$ such that $f(a) \neq b$ for every $a \in A$.

For example, consider the function $f : \mathbb{R} \to \mathbb{R}$ defined by $f(x) = x^2$. Then to show that f does not map onto $\mathbb{R}$, we note that $-1 \in \mathbb{R}$, but there exists no $a \in \mathbb{R}$ such that $f(a) = -1$. You should be able to show that f does map $\mathbb{R}$ onto the nonnegative real numbers.

Exercise 14.8.
In Exercise 13.8, which of the functions map onto their codomain? If they do not map onto, show that carefully as well.

Solutions to Exercises

Solution to Exercise (14.4).
Carefully negating the definition, we see that to show a function $f : A \to B$ is not one-to-one, we must show that there exist x_1 and x_2 in A with $f(x_1) = f(x_2)$ and $x_1 \neq x_2$. For the particular function $f(x) = x^2$ given above, we note that the numbers 1 and -1 are both real numbers with $f(1) = f(-1)$, but $1 \neq -1$. Therefore f is not one-to-one.

Solution to Exercise (14.5).
Only the function defined in (a) is one-to-one.

Solution to Exercise (14.8).
None of the functions maps onto their codomain. The range of each function was presented in the solution to Exercise 13.8.

Problems

Problem 14.1.
For each of the following, you are asked to give an example of a function. (You should always state the domain and codomain of your function.)
(a) Give an example of a function that is both one-to-one and onto.
(b) Give an example of a function that is one-to-one, but not onto.
(c) Give an example of a function that is not one-to-one, but is onto.
(d) Give an example of a function that is neither one-to-one nor onto.

Problem 14.2.
(a) Let $f : \mathbb{R} \to \mathbb{R}$ be defined by $f(x) = x^2$. Show that f is not onto.
(b) Show that f, as defined above, maps $\mathbb{R}$ onto $\{x \in \mathbb{R} : x \geq 0\}$.
(c) Consider the function $g : \mathbb{Z} \to \mathbb{N}$ defined by $g(x) = x^2$. Is g onto?
(d) Both g and f take elements of the domain and square them. Why did we use the letter g in the previous part of this problem, rather than the letter f?

Problem 14.3.
Is the absolute value function $f : \mathbb{R} \to \mathbb{R}$ defined by $f(x) = |x|$ one-to-one? Why or why not?

Problem 14.4.
(a) Is there a one-to-one function from the set $\{1, 2, 3\}$ to the set $\{1, 3\}$? Why or why not?
(b) Is there a function mapping $\{1, 2, 3\}$ onto the set $\{1, 3\}$? Why or why not?
(c) Is there a one-to-one function mapping the open interval $(0, 2)$ to the open interval $(0, 1)$?
(d) Is there a one-to-one function mapping the set $\{x \in \mathbb{R} : x > 0\}$ to the open interval $(0, 1)$?
(e) Is there a function mapping the set $\{x \in \mathbb{R} : x > 0\}$ onto the open interval $(0, 1)$?

Problem 14.5.
Recall that the definition of one-to-one was stated in the text.
  (a) State the contrapositive of the definition.
  (b) By negating the contrapositive, complete the following defini-
      tion. A function $f : X \to Y$ is not one-to-one if . . . .

Problem 14.6.
Criticize the following definition of onto. "A function $f : X \to Y$
is onto if there exists an $x \in X$ such that for each $y \in Y$ we have
$f(x) = y$."

Problem 14.7.
Define $f : \mathbb{R} \to \mathbb{R}$ by $f(x) = 5 + (x-3)^2$.
  (a) Prove that f is not injective.
  (b) Find ran(f) and prove that your conjecture is correct.

Problem 14.8.
For each of the functions below, determine whether or not the function is one-to-one and whether or not the function is onto. If the
function is not one-to-one, give an explicit example to show what
goes wrong. If it is not onto, determine the range.
  (a) Define $f : \mathbb{R} \to \mathbb{R}$ by $f(x) = 1/(x^2 + 1)$.
  (b) Define $f : \mathbb{R} \to \mathbb{R}$ by $f(x) = \sin(x)$. (Assume familiar facts
      about the sine function.)
  (c) Define $f : \mathbb{Z} \times \mathbb{Z} \to \mathbb{Z}$ by $f(n, m) = nm$.
  (d) Define $f : \mathbb{R}^2 \times \mathbb{R}^2 \to \mathbb{R}$ by $f((x, y), (u, v)) = xu + yv$. (Do you
      recognize this function?)
  (e) Define $f : \mathbb{R}^2 \times \mathbb{R}^2 \to \mathbb{R}$ by $f((x, y), (u, v)) = \sqrt{(x-u)^2 + (y-v)^2}$.
      (Do you recognize this function?)
  (f) Let A and B be nonempty sets and let $b \in B$. Define $f : A \to
      A \times B$ by $f(a) = (a, b)$.
  (g) Let X be a nonempty set, and $\mathcal{P}(X)$ the power set of X. Define
      $f : \mathcal{P}(X) \to \mathcal{P}(X)$ by $f(A) = X \setminus A$.
  (h) Let B be a fixed proper subset of a nonempty set X. Define a
      function $f : \mathcal{P}(X) \to \mathcal{P}(X)$ by $f(A) = A \cap B$.
  (i) Let $f : \mathbb{R} \to \mathbb{R}$ be defined by
$$f(x) = \begin{cases} 2 - x & \text{if } x < 1 \\ 1/x & \text{otherwise} \end{cases}.$$

Problem 14.9.
Recall the definition of the characteristic function χ_A from Problem 13.5.

(a) Under what conditions on the set A does χ_A map X onto the set $\{0, 1\}$?

(b) Find conditions on the set A and the set X that would imply that this function is injective. Justify your answer.

Problem 14.10.
For each of the following, determine whether or not f is a function from the set A to the set B. If it is, prove that f is one-to-one, or give an example to show that f is not one-to-one. Then prove that f is onto, or give an example of an element in the codomain that is not in the range to show that f is not onto.

(a) Define $f : \mathbb{Z} \times \mathbb{Z} \to \mathbb{Z} \times \mathbb{Z}$ by $f(x, y) = (y, x)$.

(b) Define $f : \mathbb{Z} \times \mathbb{Z} \to \mathbb{Z}$ by $f(x, y) = x^2 + y^2$.

(c) Let $y \in \mathbb{R}$. Define $f : \mathbb{R} \to \mathbb{R}$ by $f(x) = y \cdot x$. (Does your answer depend on y?)

(d) Define $f : \mathcal{P}(\mathbb{Z}) \to \mathbb{Z}$ by $f(S) = \max S$.

Problem♮ 14.11.
Let $f : \mathbb{R} \to (-1, 1)$ be defined by

$$f(x) = \frac{x}{1 + |x|}.$$

Prove that f is a bijective function, mapping $\mathbb{R}$ onto the open interval $(-1, 1)$.

Problem 14.12.
Let a, b, c, and d be real numbers with $a < b$ and $c < d$. Define a bijection from the closed interval $[a, b]$ onto the closed interval $[c, d]$ and prove that your function is a bijection.

Problem 14.13.
Let $F([0, 1])$ denote the set of all real-valued functions defined on the closed interval $[0, 1]$. Define a new function $\phi : F([0, 1]) \to \mathbb{R}$ by $\phi(f) = f(0)$. Is ϕ a function from $F([0, 1])$ to $\mathbb{R}$? Is it one-to-one? Is it onto? Remember to prove all claims, and to provide examples where appropriate.

Problem 14.14.
Find a function $f : \mathbb{R} \to \mathbb{R}^+$ that is one-to-one.

Problem 14.15.
Let f be a function, $f : \mathbb{R} \to \mathbb{R}$. Define a new function $f \cdot f$ by

$$(f \cdot f)(x) = f(x) \cdot f(x).$$

Prove that $f \cdot f$ is a function. Then do the remaining parts of the problem. (You may wish to work Problem 14.14, if you haven't already done so.)

(a) Does there exist a function f for which $f \cdot f$ is one-to-one? If not, why not? If there is, what is an example?

(b) Does there exist a function f for which $f \cdot f$ maps onto $\mathbb{R}$? If not, what is $\mathrm{ran}(f \cdot f)$? Your answer will be in terms of $\mathrm{ran}(f)$.

15

CHAPTER

Inverses

Given functions $f : A \to B$ and $g : C \to D$ with ran$(f) \subseteq C$, we can define a third function called the **composite function** from A to D. (We will usually call this the **composition**, rather than the composite function.) This composition is the function $g \circ f : A \to D$ defined by $(g \circ f)(x) = g(f(x))$. So, for example, if $f : \mathbb{R} \to \mathbb{R}$ and $g : \mathbb{R} \to \mathbb{R}$ are defined by $f(x) = x^2$ and $g(x) = \sin(x)$, then $(g \circ f)(x) = g(f(x)) = g(x^2) = \sin(x^2)$. Note that the order really matters here. Using f and g as above, for example, $(f \circ g)(x) = f(g(x)) = f(\sin(x)) = (\sin(x))^2$. You can check pretty easily that these two functions are different. (Check this pretty easily.) So composition of functions is not commutative.

Consider the two functions in Figure 15.1. Here $f : A \to B$, where $A = \{a, b, c\}$ and $B = \{1, 2, 3, 4\}$, while $g : C \to D$, where $C = \{2, 3, 4, 5, 6\}$ and $D = \{\alpha, \beta, \gamma\}$. Then ran$(f) \subset C$, so the composition $g \circ f$ is defined. To determine the action of $g \circ f$ algebraically, use the definition of each. For example, $(g \circ f)(a) = g(f(a)) = g(2) = \beta$. To determine the action visually, follow the arrows, remembering that f goes first.

Take this opportunity to check that the composition of three functions satisfies the associative property. In other words, if we have three functions f, g, and h so that the composition makes sense (what

175

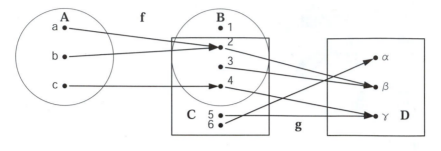

FIGURE 15.1 $g \circ f : A \to D$

would that mean?), then

$$h \circ (g \circ f) = (h \circ g) \circ f.$$

We'll use this result in this chapter.

Sometimes it is useful to "undo" the action of f. If f maps 3 to 5, we might wish to "undo" that by finding a function that takes 5 back to 3. This is most useful when we can undo the action of f on the whole range, not at just one point, because then every element ends up back where it started. For example, if f cubes all the values in its domain, we can "reverse" that action by taking the cube root. Mathematically what this means is that if $f : \mathbb{R} \to \mathbb{R}$ is defined by $f(x) = x^3$, then $g : \mathbb{R} \to \mathbb{R}$ defined by $g(x) = x^{1/3}$ satisfies two things: $(g \circ f)(x) = x$ for all $x \in \text{dom}(f)$, and $(f \circ g)(y) = y$ for all $y \in \text{dom}(g)$.

But what happens if $f : \mathbb{R} \to \mathbb{R}$ is defined by $f(x) = x^2$? If we want g to "undo" this action, then we want g to satisfy $(g \circ f)(x) = x$ for all $x \in \mathbb{R}$. But if $x = 2$, we need $(g \circ f)(2) = g(4) = 2$ and, if $x = -2$, we need $(g \circ f)(-2) = g(4) = -2$ (see Figure 15.2). What's the problem here? Well, g is not allowed to assign two different values to the number 4. So we can't do this for all functions. When can we do it? (Think first, read on later.)

Suppose that a function is bijective. Then, rather than looking in the domain and asking what x gets mapped to, we can look in the range at y and ask where it came from. Since the function is onto, y came from some x. Since the function is one-to-one, y came from exactly one x. So we can define an inverse function as follows.

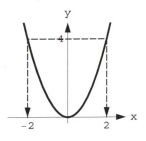

FIGURE 15.2 $f(x) = x^2$

Let $f : A \to B$ be a bijective function. The **inverse** of f is the function $f^{-1} : B \to A$ defined by

$$f^{-1}(y) = x \text{ if and only if } f(x) = y.$$

Whenever we define a function, we have to ask ourself: "Is it well-defined?" Is it? The domain is defined to be B. By definition, the value of an element of B under f^{-1} is some $x \in A$. Hence A qualifies as a codomain of f^{-1}. Now we check condition (i) of the definition of a function. Let $b \in B$. Since f is onto, there exists an element $a \in A$ such that $f(a) = b$. Hence $f^{-1}(b) = a$ is defined and property (i) holds. For property (ii) we assume that there is an element $b \in B$, and elements a and c in A such that $f^{-1}(b) = a$ and $f^{-1}(b) = c$. By the definition of f^{-1} we have $f(a) = b$ and $f(c) = b$. Hence $f(a) = f(c)$ and since f is one-to-one, $a = c$. This shows that property (ii) holds and we conclude that f^{-1} is well-defined. Note that this function is only defined in the case when f is bijective.

The discussion in the last paragraph shows that f^{-1} is indeed a function. The remainder of this chapter will be spent understanding the inverse function.

Example 15.1.
We define $f : \mathbb{R} \to \mathbb{R}$ by $f(x) = x^3 - 5$. Graph the function f. Then prove that f is one-to-one and onto. Once you have done that, decide what f^{-1} is.

"*Devising a plan.*" Assume for the moment that we know that f is bijective, so that we know that f^{-1} exists. To find f^{-1}, we use what we know: $f^{-1}(y) = x$ if and only if $f(x) = y$. Thus we must

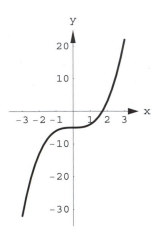

FIGURE 15.3 $f(x) = x^3 - 5$

solve $x^3 - 5 = y$ for x. Once we solve this equation, we find that $x = (y + 5)^{1/3}$. Now we are ready to solve this problem.

Proof.
We first prove that f is one-to-one. So let x_1 and x_2 be real numbers. If $f(x_1) = f(x_2)$, then $x_1^3 - 5 = x_2^3 - 5$. Therefore $x_1 = x_2$ and we may conclude that f is one-to-one.

Now we show that f is onto. Let $y \in \mathbb{R}$. Let $x = (y + 5)^{1/3}$. Then $x \in \mathbb{R}$, and $f(x) = \left((y + 5)^{1/3}\right)^3 - 5 = y$. We conclude that $y \in \mathrm{ran}(f)$. Since $f : \mathbb{R} \to \mathbb{R}$ is a well-defined function, $\mathrm{ran}(f) = \mathbb{R}$ and f maps onto $\mathbb{R}$.

We claim that $f^{-1} : \mathbb{R} \to \mathbb{R}$ is defined by $f^{-1}(y) = (y + 5)^{1/3}$. To see this, let $y \in \mathbb{R}$. We note that $f^{-1}(y) = x$ if and only if $f(x) = y$, which happens if and only if $x^3 - 5 = y$. Thus, $f^{-1}(y) = x$ if and only if $x = (y + 5)^{1/3}$ and therefore $f^{-1}(y) = (y + 5)^{1/3}$. ∎

This example brings up an important point. Students often confuse the notation f^{-1} with $1/f$. In the example above $1/f$ would be the function defined for $x \neq 5^{1/3}$ by $1/(x^3 - 5)$, while we have seen that f^{-1} is defined on all of $\mathbb{R}$ by $f^{-1}(x) = (x + 5)^{1/3}$. These two functions are really quite different! In fact, f^{-1} and $1/f$ are rarely the same. (See Project 27.6 for more information.)

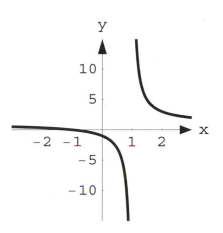

FIGURE 15.4 $f(x) = (x+1)/(x-1)$

Example 15.2.

Let $f : \mathbb{R} \setminus \{1\} \to \mathbb{R} \setminus \{1\}$ be defined by $f(x) = (x+1)/(x-1)$. (You should graph the function f, and compare it to our graph in Figure 15.4.) We'll find f^{-1}.

Before you read our solution, use Example 13.7 and Example 14.3 to check that this function is bijective and that the domain and range are also appropriate for f^{-1}.

To find an expression for f^{-1}, let $y \in \mathbb{R} \setminus \{1\}$. Then there exists (exactly one) $x \in \mathbb{R} \setminus \{1\}$ such that $f(x) = y$. Further, $f(x) = y$ if and only if $(x+1)/(x-1) = y$, and this happens if and only if $x = (y+1)/(y-1)$.

Therefore $f^{-1}(y) = x = (y+1)/(y-1)$. ○

In our examples and exercises thus far, you probably noticed us repeating the same steps: We first check that $f : A \to B$ is bijective. If it is, then we know f^{-1} exists. To find f^{-1}, we choose $y \in B$ and solve for the unique x such that $f(x) = y$. Then, by definition, $f^{-1}(y) = x$, and we are done.

Now you should be ready to do a more challenging example as an exercise.

Exercise 15.3.
Let $f : \mathbb{Z} \to \mathbb{N}$ be defined by

$$f(n) = \begin{cases} 2n & \text{if } n \geq 0 \\ -2n - 1 & \text{if } n < 0 \end{cases}.$$

We showed in Example 14.7 that this function is bijective. Find a formula for f^{-1}. (You might find it helpful to re-examine the graph of f in Figure 14.3.) $\bigcirc$

If A is a set, one very important function mapping A to itself is the identity function. So, the **identity function** i_A is the function $i_A : A \to A$ defined by $i_A(x) = x$ for all $x \in A$. You should check that i_A is well-defined, is both one-to-one and onto, and is its very own inverse. In addition, this function is easy to use. For example, if A and B are sets and f is a function such that $f : A \to B$, then $f \circ i_A = f$, while $i_B \circ f = f$.

Theorem 15.4.
Let $f : A \to B$ be a bijective function. Then
   (i) $f \circ f^{-1} = i_B$; that is, $(f \circ f^{-1})(y) = y$ for all $y \in B$.
   (ii) $f^{-1} \circ f = i_A$; that is, $(f^{-1} \circ f)(x) = x$ for all $x \in A$.
   (iii) f^{-1} is a bijective function.
   (iv) If $g : B \to A$ is a function satisfying either $f \circ g = i_B$ or $g \circ f = i_A$, then $g = f^{-1}$.

The last part of this theorem says that if we know that our function has an inverse, then f^{-1} is the one and only function satisfying the identities in (*iv*). This can come in quite handy. Consider the following.

Sometimes, as in Exercise 15.3, it is difficult to compute f^{-1}. In these cases it is nice to check your answer. Theorem 15.4 tells you one way to do so: Suppose you know that f is bijective, and you are claiming that g is the inverse. If you find that $g \circ f = i_A$ or $f \circ g = i_B$, you know you have the right answer!

We also remark here that (*i*) and (*ii*) above really follow from the definition of inverse function: $f(x) = y$ if and only if $f^{-1}(y) = x$.

Proof.
(*i*) If $y \in B$, let $z = f^{-1}(y)$. By definition $f^{-1}(y) = z$ if and only if $f(z) = y$. Therefore

$$(f \circ f^{-1})(y) = f(f^{-1}(y)) = f(z) = y.$$

(*ii*) If $x \in A$, let $z = f(x)$. By definition $f(x) = z$ if and only if $f^{-1}(z) = x$. Therefore,

$$(f^{-1} \circ f)(x) = f^{-1}(f(x)) = f^{-1}(z) = x.$$

(*iii*) We leave this for you to do in Problem 15.8.
(*iv*) We note first that $\operatorname{dom}(g) = \operatorname{dom}(f^{-1}) = B$.

First suppose that $g \circ f = i_A$. Then, using the associative property of composition and (*i*) above, we have

$$f^{-1} = i_A \circ f^{-1} = (g \circ f) \circ f^{-1} = g \circ (f \circ f^{-1}) = g \circ i_B = g.$$

In exactly the same way (except we use (*ii*) in place of (*i*)), we can show that if $f \circ g = i_B$, then $g = f^{-1}$. ∎

Before applying Theorem 15.4 make sure that you check that f really is bijective. It is one of the hypotheses, after all!

Exercise 15.5.
For each of the functions and their inverses in Example 15.1, Example 15.2, and Exercise 15.3 check that $f^{-1} \circ f = i_{\operatorname{dom}(f)}$ and $f \circ f^{-1} = i_{\operatorname{ran}(f)}$. ○

The theorem above includes the basic facts about inverses. But there are more theorems that will be useful as we move along.

Theorem 15.6.
Let $f : A \to B$ and $g : B \to C$ be bijective functions. Then $g \circ f$ is bijective and $(g \circ f)^{-1} = f^{-1} \circ g^{-1}$.

Before we begin, let's make sure we understand the function and think about what we need to prove. We note that $g \circ f : A \to C$. To show that the composition is bijective, we must show that it is one-to-one and onto. To find the inverse, if we have a guess for what it should be, we can use Theorem 15.4 (*iv*), which (in this case) says that if $g \circ f$ is bijective and h is a function such that $(h \circ (g \circ f))(x) = x$ for all x in A or $((g \circ f) \circ h)(y) = y$ for all y in B, then h must be $(g \circ f)^{-1}$. So all we need to do is think of a good candidate for h (the object we want to show is the inverse) and show it works. But the statement of the theorem gives us a candidate for h. Now that we have the plan, we can try to carry it out.

Proof.
First we'll show that the composition is one-to-one. So let $x_1, x_2 \in A$. If $(g \circ f)(x_1) = (g \circ f)(x_2)$, then $g(f(x_1)) = g(f(x_2))$. Now since g is one-to-one, $f(x_1) = f(x_2)$. But f is also one-to-one, and therefore $x_1 = x_2$, as desired.

To see that the composition is onto, let $z \in C$. Since g is onto, there exists a $y \in B$ such that $g(y) = z$. Since $y \in B$ and f is onto, there exists $x \in A$ such that $f(x) = y$. Therefore, $x \in A$ and

$$(g \circ f)(x) = g(f(x)) = g(y) = z.$$

Since $g \circ f : A \to C$, we conclude that $g \circ f$ is onto.

Now we will show that $f^{-1} \circ g^{-1}$ is the inverse of $g \circ f$ by applying (iv) of Theorem 15.4 to $g \circ f$. We just showed that $g \circ f$ is bijective. Now we check the hypotheses of part (iv) of the theorem. First, note that the domain is correct; that is, $f^{-1} \circ g^{-1} : C \to A$.

We now show that $((f^{-1} \circ g^{-1}) \circ (g \circ f))(z) = z$ for all $z \in A$. By (ii) of Theorem 15.4 applied twice (as well as the associative property of composition), for every $z \in A$ we have

$$((f^{-1} \circ g^{-1}) \circ (g \circ f))(z) = f^{-1}(g^{-1}(g(f(z)))) = f^{-1}(f(z)) = z.$$

Using (iv) of Theorem 15.4 we may conclude that $f^{-1} \circ g^{-1} = (g \circ f)^{-1}$. ∎

Remember that Pólya suggests that after solving a problem, we should look back and see whether we can use the result or the method to solve a different problem. Here's a good chance to try that out: Use the proof above to establish the following.

Theorem 15.7.
Let $f : A \to B$ and $g : B \to C$ be functions.
  (i) If f and g are one-to-one, then $g \circ f$ is one-to-one.

(ii) If f and g are onto, then g ∘ f is onto.

The converses of the two statements in the theorem above are not true. However, two corresponding weaker statements can be made. In addition, part *(iii)* of Theorem 15.8 provides a useful characterization of the inverse.

Theorem 15.8.
Let $f : A \to B$ and $g : B \to C$ be functions.
  (i) If $g \circ f$ is onto, then g is onto.
  (ii) If $g \circ f$ is one-to-one, then f is one-to-one.
  (iii) Suppose now that $f : A \to B$ and $g : B \to A$. If $f \circ g = i_B$ and $g \circ f = i_A$, then $g = f^{-1}$.

How does *(iii)* in Theorem 15.8 differ from part *(iv)* in Theorem 15.4? Well, both are implications, but the antecedent in one is a disjunction and the antecedent in the other is a conjunction. In addition, in Theorem 15.8, we do not assume that either of the functions are bijective. You will need to show that the conditions in *(iii)* imply that f and g are, in fact, bijective. If you already know that one of your functions f and g are bijective, Theorem 15.4 will usually be easier to use than Theorem 15.8.

Exercise 15.9.
Prove Theorem 15.8   ○

Solutions to Exercises

Solution to Exercise (15.3).
This problem only asks for a formula for f^{-1}, which we will give here. You need to think about how we obtained this formula. You should check that $f^{-1} : \mathbb{N} \to \mathbb{Z}$ defined by

$$f^{-1}(m) = \begin{cases} m/2 & \text{if } m \text{ is even} \\ -(m+1)/2 & \text{if } m \text{ is odd} \end{cases}$$

really is the inverse of f.

Solution to Exercise (15.9).
(i) If $c \in C$, then the fact that $g \circ f$ is onto implies that there exists $a \in A$ such that $(g \circ f)(a) = c$. Therefore $g(f(a)) = c$. Since $f(a) \in B$, we have shown that there is an element $b = f(a)$ in B such that $g(b) = c$. Since $g : B \to C$, we conclude that g is onto.

(ii) If $a_1, a_2 \in A$, and $f(a_1) = f(a_2)$, then $g(f(a_1)) = g(f(a_2))$. Therefore, $(g \circ f)(a_1) = (g \circ f)(a_2)$. Since $g \circ f$ is one-to-one, $a_1 = a_2$ and f is one-to-one, as desired.

(iii) Since $g \circ f$ is one-to-one, (ii) implies that f is one-to-one. Since $f \circ g$ is onto, (i) implies that f is onto. Thus f is bijective. Consequently, (iii) follows from Theorem 15.4 (iv). $\square$

Problems

In all the problems below, the sets A and B are nonempty.

Problem 15.1.
Find the compositions $f \circ g$ and $g \circ f$ assuming the domain of each is the largest set of real numbers for which the functions f, g, $f \circ g$, and $g \circ f$ make sense. In your solution to each of the following, give the compositions and the corresponding domain and range:
(a) $f(x) = 1/(1 + x)$, $g(x) = x^2$;
(b) $f(x) = x^2$, $g(x) = \sqrt{x}$ (simplify this one);
(c) $f(x) = 1/x$, $g(x) = x^2 + 1$;
(d) $f(x) = |x|$, $g(x) = f(x)$.

Problem 15.2.
Let $f : \mathbb{R} \to \mathbb{R}$ be defined by $f(x) = x^3 + 4$. Use Theorem 15.8 to show that if $g : \mathbb{R} \to \mathbb{R}$ is defined by $g(x) = (x - 4)^{1/3}$, then $g = f^{-1}$.

Problem 15.3.
Let $f : \mathbb{R}^2 \to \mathbb{R}$ be defined by $f(x, y) = x + y$. Prove that there is no function $g : \mathbb{R} \to \mathbb{R}^2$ such that $g \circ f = i_{\mathbb{R}^2}$.

Problem 15.4.
Let $f : \mathbb{R} \to \mathbb{R}^2$ be defined by $f(x) = (x, 0)$. Show that there is no function $g : \mathbb{R}^2 \to \mathbb{R}$ such that $f \circ g = i_{\mathbb{R}^2}$.

Problem 15.5.
For each of the following, find the range of f. If possible, find f^{-1} : $\operatorname{ran}(f) \to \operatorname{dom}(f)$. If it is not possible, explain why it is not possible for:
  (a) the function f defined on the nonzero real numbers by $f(x) = 1/x$;
  (b) the function $f : \mathbb{R}^2 \to \mathbb{R}$ defined by $f(x, y) = x + y$;
  (c) the function $f : \mathbb{R}^2 \to \mathbb{R}^2$ defined by $f(x, y) = (y, x)$;
  (d) the function $f : \mathbb{R} \to \mathbb{R}$ defined by $f(x) = \sin x$;
  (e) the function f defined by $f(x) = \tan x$, where $-\pi/2 < x < \pi/2$.

Problem 15.6.
The functions $f : \mathbb{R} \setminus \{-2\} \to \mathbb{R} \setminus \{1\}$ and $g : \mathbb{R} \setminus \{1\} \to \mathbb{R} \setminus \{-2\}$ defined by

$$f(x) = \frac{x - 3}{x + 2} \text{ and } g(x) = \frac{3 + 2x}{1 - x}$$

are well-defined functions (you need not check this).
  (a) Calculate $f \circ g$ and $g \circ f$.
  (b) What can you conclude about f and g from your result in part (a)? If you use a theorem, give a reference.

Problem 15.7.
  (a) If possible, find examples of functions $f : A \to B$ and $g : B \to A$ such that $f \circ g = i_B$ when:
      (i) $A = \{1, 2, 3\}$, $B = \{4, 5\}$;
      (ii) $A = \{1, 2\}$, $B = \{4, 5\}$;
      (iii) $A = \{1, 2, 3\}$, $B = \{4, 5, 6, 7\}$.
      Draw diagrams of A and B in each case above.
  (b) Give an example of sets A and B, and functions $f : A \to B$ and $g : B \to A$ such that $f \circ g = i_B$, but $g \circ f \neq i_A$. (Thus the existence of a function g such that $f \circ g = i_B$ is *not* enough to conclude that f has an inverse!) Why doesn't this contradict Theorem 15.4, part *(iv)*?

(c) Give an example of sets A and B, and functions $f : A \to B$ and $g : B \to A$ such that $g \circ f = i_A$, but $f \circ g \neq i_B$. (Thus the existence of a function g such that $g \circ f = i_A$ is not enough to conclude that f has an inverse!) Why doesn't this contradict Theorem 15.4, part (iv)?

(d) Let A and B be two sets, and let $f : A \to B$ be a function. Assume further that there exists a function $g : B \to A$ such that $f \circ g = i_B$. Must f be one-to-one? onto?

(e) Looking over your work above, what should be your strategy in solving a question like (d) above? Whatever you decide, use it to solve the following: Let f and g be as above and suppose $g \circ f = i_A$. Must f be one-to-one? onto?

Problem 15.8.
Let f be a bijective function. Prove part (iii) of Theorem 15.4 and show that $(f^{-1})^{-1} = f$.

Problem 15.9.
(a) Give an example of a function $f : A \to A$ such that $f \neq i_A$, but $f \circ f = i_A$. Must such a function f be one-to-one? onto?

(b) Give an example of a nonzero function $f : \mathbb{R} \to \mathbb{R}$ such that $(f \circ f)(x) = 0$ for all $x \in \mathbb{R}$. Can such a function be one-to-one? onto?

Problem 15.10.
Let $f : A \to A$ be a function. Suppose that the composition $f \circ f$ is a bijection. Must such a function f be a bijection? (Prove this or give a counterexample.)

Problem 15.11.
Suppose that $f : A \to B$ and g_1 and g_2 are functions from B to A such that $f \circ g_1 = f \circ g_2$. Show that if f is bijective, then $g_1 = g_2$. If $g_1 \circ f = g_2 \circ f$ and f is bijective, must $g_1 = g_2$?

Problem 15.12.
Let $f : A \to A$ be a function. Define a relation on A by $a \sim b$ if and only if $f(a) = f(b)$. Is this an equivalence relation? If f is one-to-one, what is the equivalence class of a point $a \in A$?

Problem 15.13.
Let $f : A \to A$ be a function. Define a relation on A by $a \sim b$ if and only if $f(a) = b$. Is this an equivalence relation for an arbitrary function f? If not, is there a function for which it is an equivalence relation?

Problem♮ 15.14.
Let A, B, C, and D be nonempty sets. Let $f : A \to B$ and $g : C \to D$ be functions.
  (a) Prove that if f and g are one-to-one, then $H : A \times C \to B \times D$ defined by

$$H(a, c) = (f(a), g(c))$$

  is a one-to-one function. (Check that it is one-to-one and a function.)
  (b) Prove that if f and g are onto, then H is also onto.

Problem♮ 15.15.
Let A, B, C, and D be nonempty sets. Let $f : A \to B$ and $g : C \to D$ be functions. Consider H defined on $A \cup C$ by

$$H(x) = \begin{cases} f(x) & \text{if } x \in A \\ g(x) & \text{if } x \in C \end{cases}.$$

Show that there exist sets A, B, C, and D for which H is *not* a function, but there also exist such sets for which H is a function. What conditions can we place on A and C that assure us that H is a function?

Problem 15.16.
Let $a \in \mathbb{R}$ with $|a| < 1$. Define f on the set $\{x \in \mathbb{R} : |x| < 1\}$ by

$$f(x) = \frac{a - x}{1 - ax}.$$

  (a) Show that the range of f is contained in the set $\{x \in \mathbb{R} : |x| < 1\}$.
  (b) Does f map onto the set $\{x \in \mathbb{R} : |x| < 1\}$?
  (c) Prove that f is one-to-one.
  (d) Compute $f \circ f$.
  (e) Find f^{-1}.

Problem 15.17.
Let $\mathbb{R}[x]$ denote the set of all polynomials with real coefficients. (See Problem 10.8.)
 (a) Define a function f on $\mathbb{R}[x]$ by $f(p) = p(0)$. What is the range of f? Is f one-to-one?
 (b) Define a function g on the nonzero polynomials in $\mathbb{R}[x]$ by $g(p) =$ degree of p. Is g a function? Is it one-to-one? What is the range of g?
 (c) Recall that a value z is a root of a polynomial p if $p(z) = 0$. Define F on $\mathbb{R}[x]$ by $F(p) = $ a root of p. Is F a function? Why or why not?
 (d) Define h on $\mathbb{R}[x]$ by $(h(p))(x) = xp(x)$. Is h a function? If so, is it one-to-one? What is the range of h?
 (e) Define $k : \mathbb{R}[x] \to \mathbb{R}[x]$ by $k(p) = p \circ p$. Show that k is neither one-to-one nor onto.

Problem 15.18.
For each part give examples of functions $f : A \to B$ and $g : B \to C$ satisfying the stated conditions.
 (a) The composition $g \circ f$ is onto, but f is not onto.
 (b) The composition $g \circ f$ is one-to-one, but g is not one-to-one.

Problem 15.19.
Let $f : \mathbb{R} \to \mathbb{R}$ be a function such that f is onto and $f \circ f \circ f = f$. Prove that f is bijective.

Problem 15.20.
In this problem, we look at a function called the restriction function, which we now define.
 If $f : A \to B$ is a function, and $A_1 \subset A$, we define another function $F : A_1 \to B$ by $F(a) = f(a)$ for all $a \in A_1$. This function F is called the **restriction** of f to A_1 and is usually denoted $f|_{A_1}$. We now turn to the problem:
 (a) Prove that if f is one-to-one, then $f|_{A_1}$ is one-to-one.
 (b) Prove that if $f|_{A_1}$ is onto, then f is onto.

Problem 15.21.
Suppose that A, B, C, and D are nonempty sets with $B \subseteq C$ and $D \subseteq A$. Suppose that both functions $f : A \to B$ and $g : C \to D$ are onto and $f \circ g \circ f = f$. (Note that the compositions $g \circ f$ and $f \circ g$ are both defined.) Refer to Problem 15.20 for the definitions of $(f \circ g)|_B$ and $(g \circ f)|_D$.

 (a) Show that $(f \circ g)|_B$ is one-to-one.
 (b) Give an example to show that $(g \circ f)|_D$ is not necessarily one-to-one.

16

CHAPTER

Images and Inverse Images

In the last chapter, we looked at where points in the domain are mapped to under a function f and where points in the range come from under f. But sometimes we need to look at where f maps a whole set, or where an entire set comes from. So here are two definitions that are waiting to be understood.

Let $f : X \to Y$ be a function and let $A \subseteq X$. Then the **image** of A under f is the set

$$f(A) = \{f(a) : a \in A\}.$$

Note that $f(A)$ is the notation we use for this set, and that this set is a subset of Y. In "street talk" the image of A under f is where the elements of A were taken by f.

Exercise 16.1.
It's good to start small. So let's begin with $A = \{1, 2, 4\}$ and $B = \{-1, 1, -2, 3\}$. Find each of the requested images under the function $f : \mathbb{R} \to \mathbb{R}$ defined by $f(x) = x^2$.
  (a) What is $f(A)$?
  (b) What is $f(B)$?
  (c) What is $f(A \cap B)$?
  (d) What is $f(A) \cap f(B)$?

We'll solve (a) for you here, so you can see what we are asking you to do. We claim that $f(A) = \{1, 4, 16\}$. To see this, we use the definition:

$$f(A) = \{f(a) : a \in A\} = \{f(1), f(2), f(4)\} = \{1, 4, 16\}. \qquad \bigcirc$$

Small sets are easier because you can often list the values, just as we did above. This won't be possible, in general, as you will see below.

We are also interested in where sets in the codomain come from. This is called the inverse image of a set (because we are going backwards) and there is one unfortunate thing about it: the notation involves the symbol f^{-1}, which we have used only when f is bijective. Well, here f may not be bijective, and therefore, f^{-1} may not be a function. Though this may be confusing at first, this is generally agreed upon notation and you (the reader) must check carefully on the context. Having said all that, we now define the inverse image.

Let $f : X \rightarrow Y$ be a function and let $B \subseteq Y$. Then the **inverse image** of B under f is the set $f^{-1}(B)$ defined by

$$f^{-1}(B) = \{x \in X : f(x) \in B\}.$$

In other words, the inverse image of B is *the subset of X* consisting of all the elements in the domain that get mapped into B. Note that when f is *not* bijective, the notation $f^{-1}(y)$ makes no sense (why?). If you want to talk about the inverse image of a set with just one element, say so by writing $f^{-1}(\{y\})$. (You may find texts in which the authors use the notation $f^{-1}(y)$, but we find that it often introduces unnecessary confusion.)

Exercise 16.2.
Let $f : \mathbb{R} \rightarrow \mathbb{R}$ be defined by $f(x) = x^2$. Find:
  (a) $f^{-1}(\{4\})$;
  (b) $f^{-1}(\{1, 2, 4\})$;
  (c) $f^{-1}(f(A))$, where $A = \{1, 2\}$.

Again, we will do (a) here, so you can see what we are asking you to do. By definition,

$$f^{-1}(\{4\}) = \{x \in \mathbb{R} : f(x) \in \{4\}\} = \{x \in \mathbb{R} : f(x) = 4\}.$$

Replacing f by what it equals, we have

$$f^{-1}(\{4\}) = \{x \in \mathbb{R} : x^2 = 4\} = \{-2, 2\}. \qquad \bigcirc$$

Since the sets above are small, we can list all the elements. We ask that you now check your understanding with more challenging sets, but still using the same function as in the previous exercises.

By carefully writing out the definitions of the sets in Exercise 16.3, it is possible to guess what the answers are. We provide rigorous proofs for several parts at the end of this chapter. If you wish to try them yourself first (which you are certainly encouraged to do), make sure that you work from the inside out on parts (e)–(h). So in part (e), for example, first find $f([0,1])$ (which works just like (a)) and call that set A. Then find $f^{-1}(A)$ (which works just like (d)).

Exercise 16.3.
Let $f : \mathbb{R} \to \mathbb{R}$ be defined by $f(x) = x^2$. Find:
  (a) $f([-1,1])$;
  (b) $f(\mathbb{Z})$;
  (c) $f^{-1}(\mathbb{N})$;
  (d) $f^{-1}([-1,0])$;
  (e) $f^{-1}(f([0,1]))$;
  (f) $f(f^{-1}([-1,0]))$;
  (g) $f^{-1}([0,1] \cup [2,4])$;
  (h) $f([0,1] \cap [-1,0])$;
  (i) $f([0,1]) \cap f([-1,0])$. $\qquad \bigcirc$

Your experience with concrete sets will help you work with abstract sets.

Exercise 16.4.
Looking back at the examples in the exercises above, decide which of the following you think are true for all functions $f : X \to Y$, all subsets A and B of X, and all subsets C and D of Y:
  (a) $f(f^{-1}(C)) = C$;
  (b) $f^{-1}(f(A)) = A$;
  (c) $f(A \cap B) = f(A) \cap f(B)$;

(d) $f(A) = f(B)$ implies that $A = B$;
(e) $f^{-1}(C) = f^{-1}(D)$ implies that $C = D$.   ○

All of the statements above may look reasonable, yet they are all false. Nevertheless, there are many similar statements that are true. You can prove them all with the tools you have developed at this point. To emphasize the accepted writing techniques, we provide an example below.

Theorem 16.5.
Let $f : X \to Y$ and let A_1 and A_2 be subsets of X. Then $f(A_1 \cap A_2) \subseteq f(A_1) \cap f(A_2)$.

"Understanding the problem." Remember that you won't get any-where if you don't know the definitions. So we need to figure out what the sets in the statement are. We begin by making sure we know what $f(A_1 \cap A_2)$, $f(A_1)$, and $f(A_2)$ are. First, $f(A) = \{f(x) : x \in A\}$. So that should make it pretty clear. Things in $f(A_1 \cap A_2)$ look like $f(x)$ where $x \in A_1 \cap A_2$. Now it should occur to you that you must write out what it means to be in $f(A_1) \cap f(A_2)$. Once you have done that, you have done the preliminaries.
"Devising a plan." When we worked with sets with a special form (like the Cartesian product of two sets) we emphasized that if we never used the special form of the elements, we would most likely never prove the desired result. The same is true here—if we never use the fact that the elements have the form $f(x)$ where $x \in A_1 \cap A_2$ we shouldn't expect to be able to prove the result. The next step is to note that what we want to do is to show that one set is contained in another set. We know how to do that, too. So our plan is to start with an element in the set on the left side, use the special form of this set and show the element is in the set on the right. As we *"carry out our plan,"* note how quickly we move to the special form of the element.

Proof.
If $y \in f(A_1 \cap A_2)$, then $y = f(x)$ for some $x \in A_1 \cap A_2$. Since $x \in A_1 \cap A_2$, we have $x \in A_1$ and $x \in A_2$. Since $x \in A_1$ and $y = f(x)$, we see that

$y \in f(A_1)$. Similarly, since $x \in A_2$ and $y = f(x)$, we see that $y \in f(A_2)$. Therefore $y \in f(A_1) \cap f(A_2)$. Thus $f(A_1 \cap A_2) \subseteq f(A_1) \cap f(A_2)$. ∎

Exercise 16.6.

We already have an example to show that, with the notation from the theorem above, we need not have $f(A_1 \cap A_2) = f(A_1) \cap f(A_2)$. But what is wrong with the following proof of this "non-fact"?

Not a proof.

It follows from Theorem 16.5 that $f(A_1 \cap A_2) \subseteq f(A_1) \cap f(A_2)$. To show the reverse set inclusion, we let $y \in f(A_1) \cap f(A_2)$. By definition of intersection, $y \in f(A_1)$ and $y \in f(A_2)$. Therefore, $y = f(x)$ for some x in A_1 and $y = f(x)$ for some $x \in A_2$. Since $x \in A_1$ and $x \in A_2$, we see that $x \in A_1 \cap A_2$. Thus $y = f(x)$ where $x \in A_1 \cap A_2$, so $y \in f(A_1 \cap A_2)$. This proves that $f(A_1) \cap f(A_2) \subseteq f(A_1 \cap A_2)$, and the non-fact is established! ☐

We know there's something wrong above since the assertion isn't always true. But it isn't always false either. Find the error and see if you can think of another hypothesis we might place on f that would help us to determine the functions for which the assertion is true. ○

The next theorem is one you will use repeatedly. You really can do all the proofs yourself.

Theorem 16.7.

Let $f : X \to Y$. Let $A, A_1,$ and A_2 be subsets of X and $B, B_1,$ and B_2 subsets of Y. Then

1. *if $A_1 \subseteq A_2$, then $f(A_1) \subseteq f(A_2)$;*
2. $f(A_1 \cup A_2) = f(A_1) \cup f(A_2)$;
3. $f(A_1 \cap A_2) \subseteq f(A_1) \cap f(A_2)$;
4. *in general, $f(X \setminus A) \neq Y \setminus f(A)$;*
5. *if $B_1 \subseteq B_2$, then $f^{-1}(B_1) \subseteq f^{-1}(B_2)$;*
6. $f^{-1}(B_1 \cap B_2) = f^{-1}(B_1) \cap f^{-1}(B_2)$;
7. $f^{-1}(B_1 \cup B_2) = f^{-1}(B_1) \cup f^{-1}(B_2)$;
8. $f^{-1}(Y \setminus B) = X \setminus f^{-1}(B)$;
9. $A \subseteq f^{-1}(f(A))$;
10. $f(f^{-1}(B)) \subseteq B$.

We have already presented a proof of (3) in Theorem 16.5, and we will provide a proof of (9) in Example 16.8. The other parts of the theorem are left to the reader (that's you) in the problems. Remember that before beginning the proof of each part you must make sure you know what the left-hand side is, and what the right-hand side is. We suggest that you write out the element definition of both sides carefully (as we do in Example 16.8 below), and then show that appropriate relations hold using acceptable mathematical and writing techniques.

If additional conditions are placed on the function f, then some of the conclusions in Theorem 16.7 can be strengthened. We look at such a case in the following example. Some of the problems will ask you to consider similar restrictions.

Example 16.8.
We will prove part 9 of Theorem 16.7. Then we will show that the inclusion is, in general, proper. We conclude this example by showing that if f is required to be one-to-one, then the two sets are actually equal.

(a) First we prove that if $f : X \to Y$ and $A \subseteq X$, then $A \subseteq f^{-1}(f(A))$. Before we begin, we note that the right side is a bit complicated. Let's make sure we understand it: Since $f^{-1}(B) = \{x \in X : f(x) \in B\}$ replacing B by $f(A)$, we see that $f^{-1}(f(A)) = \{x \in X : f(x) \in f(A)\}$. So we must show that if $z \in A$, then $z \in \{x \in X : f(x) \in f(A)\}$; in other words, we must show that $z \in X$ and $f(z) \in f(A)$.

Proof.
If $z \in A$, then since $A \subseteq X$, we know that $z \in X$. By the definition of $f(A)$, we have $f(z) \in f(A)$. Consequently, $z \in f^{-1}(f(A))$, and $A \subseteq f^{-1}(f(A))$. ∎

(b) Figure 16.1 indicates why, for an arbitrary function and an arbitrary set A, we cannot expect that the two sets A and $f^{-1}(f(A))$ are equal.

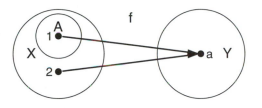

FIGURE 16.1 $f : \{1, 2\} \to \{a\}$

The diagram shows that if we let $A = \{1\}$ and define $f : \{1, 2\} \to \{a\}$, by $f(1) = f(2) = a$, then $A = \{1\}$ while $f^{-1}(f(A)) = f^{-1}(f(\{1\})) = f^{-1}(\{a\}) = \{1, 2\}$. Thus $A \neq f^{-1}(f(A))$.

(c) However, if the function $f : X \to Y$ is one-to-one and $A \subseteq X$, then $A = f^{-1}(f(A))$.

Proof.
The inclusion $A \subseteq f^{-1}(f(A))$ is proven in (a) above. For the reverse inclusion, suppose $z \in f^{-1}(f(A))$. Then $z \in X$ and $f(z) \in f(A)$. Thus, there exists $x \in A$ such that $f(z) = f(x)$. Now f is one-to-one and so $z = x$. But $x \in A$, so $z \in A$. Hence $f^{-1}(f(A)) \subseteq A$, and we conclude that the two sets are equal. ∎

Solutions to Exercises

Solution to Exercise (16.1).
You should be able to check that:
(b) $f(B) = \{1, 4, 9\}$;
(c) $f(A \cap B) = \{1\}$;
(d) $f(A) \cap f(B) = \{1, 4\}$.

Solution to Exercise (16.2).
You should be able to check that:
(b) $f^{-1}(\{1, 2, 4\}) = \{-1, 1, -\sqrt{2}, \sqrt{2}, -2, 2\}$;
(c) $f^{-1}(f(A)) = f^{-1}(\{1, 4\}) = \{-1, 1, -2, 2\}$.

Solution to Exercise (16.3).
We give complete solutions to (a), (d), (f), and (h) followed by answers to (b), (c), (e), (g), and (i).
  (a) We claim that $f([-1,1]) = [0,1]$. To see this, let $y \in f([-1,1])$. By definition of the image,

$$f([-1,1]) = \{f(x) : x \in [-1,1]\} = \{x^2 : x \in [-1,1]\}.$$

  So there exists $x \in [-1,1]$ such that $y = x^2$. Since $x \in [-1,1]$, we know that $0 \le x^2 \le 1$. Therefore, $y \in [0,1]$, and $f([-1,1]) \subseteq [0,1]$.
  Now suppose that $y \in [0,1]$. Let $x = \sqrt{y}$. Then $x \in [0,1] \subset [-1,1]$, and $f(x) = x^2 = (\sqrt{y})^2 = y$. Therefore, there exists $x \in [-1,1]$ such that $y = f(x)$ and $y \in f([-1,1])$. So $[0,1] \subseteq f([-1,1])$, and we conclude that the two sets are equal.
  (d) We claim that $f^{-1}([-1,0]) = \{0\}$. To see this, let $x \in f^{-1}([-1,0])$. By definition,

$$f^{-1}([-1,0]) = \{x \in \mathbb{R} : f(x) \in [-1,0]\} = \{x \in \mathbb{R} : x^2 \in [-1,0]\}.$$

  Therefore $x \in f^{-1}([-1,0])$ implies that $x^2 \in [-1,0]$. This is only possible if $x = 0$. Therefore $f^{-1}([-1,0]) \subseteq \{0\}$. Now suppose that $x \in \{0\}$. Then $f(x) = f(0) = 0$. Therefore, $f(x) \in [-1,0]$, and $x \in f^{-1}([-1,0])$. Consequently, $\{0\} \subseteq f^{-1}([-1,0])$, and we conclude that $f^{-1}([-1,0]) = \{0\}$.
  (f) We claim that $f(f^{-1}([-1,0])) = \{0\}$. By (d), we know $f^{-1}([-1,0]) = \{0\}$. Therefore, we need to find

$$f(f^{-1}([-1,0])) = f(\{0\}).$$

  Thus

$$f(f^{-1}([-1,0])) = f(\{0\}) = \{f(0)\} = \{0\},$$

  as desired.
  (h) We work from the inside out. Thus,

$$f([0,1] \cap [-1,0]) = f(\{0\}) = \{f(0)\} = \{0\}.$$

The answer to (b) is $\{z^2 : z \in \mathbb{N}\}$; to (c) is $\{\sqrt{n} : n \in \mathbb{N}\} \cup \{-\sqrt{n} : n \in \mathbb{N}\}$; to (e) is $[-1,1]$; to (g) is $[-1,1] \cup [-2,-\sqrt{2}] \cup [\sqrt{2},2]$; and to (i) is $[0,1]$.

There are many possible ways to solve these problems. Though each problem seems to be different, what remains the same in every problem is that you need to understand the notation and the definitions.

Solution to Exercise (16.6).
In "not a proof" the conclusion "$y = f(x)$ for some x in A_1 and A_2" is not justified. From the assertion that $y \in f(A_1)$ and $y \in f(A_2)$, we may only conclude that there exists an $x_1 \in A_1$ such that $y = f(x_1)$, and there exists an $x_2 \in A_2$ such that $y = f(x_2)$. We may not conclude that $x_1 = x_2$. Indeed, Exercise 16.1 shows that there are cases where neither of the two elements is in the intersection.

However, the following is true. Let $f : X \to Y$ be an injective function, and let A_1 and A_2 be subsets of X. Then $f(A_1 \cap A_2) = f(A_1) \cap f(A_2)$. Why? Well, we know that $f(x_1) = y = f(x_2)$ and we now know that f is injective. Thus, $x_1 = x_2$, so $x_1 \in A_1 \cap A_2$. The rest of the "not a proof" is valid.

Spotlight: Minimum or Infimum?

If you've ever forgotten to show that the infimum of a set was the minimum, this historical example should show you that it can happen to the best of us.

Suppose you know the temperature on the surface of the earth (because people on the surface measured it) and you want the temperature on the inside (but no one can get to the place to measure it). How do you get the temperature? This kind of question interested several famous mathematicians. In fact, there's plenty of mathematical research done today that is related to this problem.

This problem is known as the Dirichlet problem. It can be studied in more generality, but we'll stick to looking at functions defined on a sphere. To understand the statement, you need to have studied several variable calculus. At the end of this spotlight, we will state the problem on the sphere, along with the references, for those of you who have the background.

The solution to this problem used something called the "Dirichlet principle." The idea of the principle was to look at a collection of certain integrals with the region of integration fixed, but with different integrands. It was first argued that the values of the integrals were bounded below (and therefore, as we have learned, there was an infimum). From there, the mathematicians assumed that for one of the functions, the integral was the minimum. This principle was used by many excellent mathematicians: by George Green, Georg Friedrich Bernhard Riemann, Sir William Thomson (also known as Lord Kelvin), and others. In 1870, Karl Theodor Wilhelm Weierstrass presented an important paper about the validity of this argument. Even the title of his article "Über das sogenannte Dirichlet'sche Princip" (On the so-called Dirichlet principle) is enough to show what Weierstrass thought of the principle [82]. He began his paper by reconstructing Dirichlet's argument. He then explained that though an expression may have a lower bound that we can get arbitrarily close to, we may never actually reach it. Weierstrass concluded his paper with an example showing how this might happen. Almost thirty years later, David Hilbert supplied a proof of the principle for certain cases when he presented what he called the "resuscitation" of the Dirichlet principle [68, p. 67].

It may seem odd that mathematicians of this calibre would use an unproven principle. There are two things to remember. First, the principle was supported on physical grounds. Second, rigor was still being introduced to mathematics. In spite of its unusual history, the Dirichlet principle served an important purpose. In Kline's words [48, p. 704] "Had the progress made with the use of the principle awaited Hilbert's work, a large segment of nineteenth-century work on potential theory and function theory would have been lost."

More information on this is available in [48], [26], and [58]. Another criticism of the Dirichlet principle, from a different point of view, was published by Friedrich Prym. More information about this can be found in [69]. The statement of the problem in $\mathbb{R}^3$ is the following: Let f be a continuous real-valued function on the sphere of radius one (the unit sphere). A real-valued function g is called harmonic on the open unit ball ($\{(x, y, z) \in \mathbb{R}^3 : x^2 + y^2 + z^2 < 1\}$) if g

has continuous second partial derivatives satisfying

$$g_{xx} + g_{yy} + g_{zz} = 0$$

throughout the ball. The question is: Does there exist a function F that is continuous on the closed ball of radius one, equal to f on the unit sphere, and harmonic on the open unit ball?

Problems

Problem 16.1.
Recall that $[a, b]$ denotes the closed interval from a to b, while (a, b) denotes the open interval. For the function $f : \mathbb{R} \to \mathbb{R}$ defined by $f(x) = 3x - 1$, find:
  (a) $f((0, 1))$;
  (b) $f((a, b))$, where $a, b \in \mathbb{R}$ and $a < b$;
  (c) $f^{-1}((-2, -1))$;
  (d) $f^{-1}((a, b))$, where $a, b \in \mathbb{R}$ and $a < b$.

Problem 16.2.
For the function $f : \mathbb{R} \to \mathbb{R}$ defined by $f(x) = 2x^2$, find:
  (a) $f((0, 1))$;
  (b) $f((-1, 3))$;
  (c) and (in general) $f((a, b))$, where $a, b \in \mathbb{R}$ and $a < b$;
  (d) $f^{-1}((-2, -1))$;
  (e) $f^{-1}((0, 2))$;
  (f) and (in general) $f^{-1}((a, b))$, where $a, b \in \mathbb{R}$ and $a < b$.
Actually, we are really only interested in your answers to (c) and (f). So why did you have to work all the other parts?

Problem 16.3.
For the function $f : \mathbb{R} \to \mathbb{R}$ defined by $f(x) = |x|$, find:
  (a) $f((-1, 1))$;
  (b) $f(\{-1, 1\})$;
  (c) $f^{-1}(\{1\})$;
  (d) $f^{-1}([-1, 0))$;
  (e) $f^{-1}(f([0, 1]))$.

Problem 16.4.
Consider the function $\chi_{(0,1)} : \mathbb{R} \to \mathbb{R}$ (this is the characteristic function as defined in Problem 13.5). Find:
  (a) $\chi_{(0,1)}((0,1))$;
  (b) $\chi_{(0,1)}((-1,3))$;
  (c) and (in general) $\chi_{(0,1)}((a,b))$, where $a, b \in \mathbb{R}$ and $a < b$;
  (d) $\chi_{(0,1)}^{-1}((-2,-1))$;
  (e) $\chi_{(0,1)}^{-1}((0,2))$;
  (f) and (in general) $\chi_{(0,1)}^{-1}((a,b))$, where $a, b \in \mathbb{R}$ and $a < b$.
Actually, we are really only interested in your answers to (c) and (f). So why did you have to work all the other parts?

Problem 16.5.
We denote the characteristic function of $\mathbb{Z}$ in $\mathbb{R}$ by $\chi_\mathbb{Z} : \mathbb{R} \to \mathbb{R}$ (see Problem 13.5 for the definition). In each case below, start by writing out the definition for the particular set and function. Then write the solution to each of the following in as simple a form as possible:
  (a) $\chi_\mathbb{Z}(\mathbb{Z}^+)$;
  (b) $\chi_\mathbb{Z}^{-1}(\mathbb{Z}^+)$;
  (c) $\chi_\mathbb{Z}(\chi_\mathbb{Z}^{-1}(\mathbb{Z}^+))$;
  (d) $\chi_\mathbb{Z}^{-1}(\chi_\mathbb{Z}(\mathbb{Z}^+))$.

Problem 16.6.
Let $f : \mathbb{R} \to \mathbb{R}$ be defined by $f(x) = x^4 + 1$.
  (a) Make a careful graph of f.
  (b) Using your graph, show how you can guess $f([0,2])$.
  (c) Show that your guess for $f([0,2])$ is correct.
  (d) Use your graph to find $f^{-1}([2,17])$.
  (e) Show that your guess for $f^{-1}([2,17])$ is correct.

Problem 16.7.
Let p and q be two polynomials of degree two with real coefficients. (See Problem 10.8 for definitions.) Suppose $p^{-1}(\{0\}) = q^{-1}(\{0\})$.
  (a) Give an example of such p and q, with $p \neq q$.
  (b) Suppose that $p^{-1}(\{0\}) = \{0,1\} = q^{-1}(\{0\})$. Must $p = q$? Either prove this or give a counterexample.

Problem 16.8.
Let $f : \mathbb{Z} \to \mathbb{N}$ be defined by

$$f(n) = \begin{cases} -2n & \text{if } n \leq 0 \\ 2n - 1 & \text{if } n > 0 \end{cases} .$$

Recall that $2\mathbb{Z} = \{2n : n \in \mathbb{Z}\}$. Find $f(2\mathbb{Z})$ and prove that your answer is correct.

Problem 16.9.
Prove Theorem 16.7, part 1.

Problem 16.10.
Using Theorem 16.7 part 1, rather than element-chasing, prove Theorem 16.7 part 3.

Problem 16.11.
Prove Theorem 16.7, part 2.

Problem 16.12.
(a) Prove Theorem 16.7, part 4. We suggest the following strategy: Try to prove that the two sets are equal. If you do this carefully, you may start to wish for restrictions on f that you don't have. This should help you think of examples to show that the two sets need not be equal.
(b) If f is onto, does the statement in Theorem 16.7 part 4 become an equality? What if f is one-to-one?
(c) Show that if f is bijective, then equality holds.

Problem 16.13.
(a) Prove Theorem 16.7, part 5.
(b) In the same context, what can you conclude if $B_1 = B_2$? State your result and prove it.

Problem 16.14.
Prove Theorem 16.7, part 6.

Problem 16.15.
Prove Theorem 16.7, part 7.

Problem 16.16.
Prove Theorem 16.7, part 8.

Problem 16.17.
  (a) Prove Theorem 16.7, part 10.
  (b) Give an example to show that the two sets may not be equal.
  (c) If f is onto, must the two sets be equal?
  (d) If f is one-to-one, must the two sets be equal?

Problem 16.18.
Let $f : X \rightarrow Y$ be a function satisfying $f(A) \cap f(B) = \emptyset$ whenever A and B are sets with $A \cap B = \emptyset$.
  (a) Give an example of such a function. Prove that your example satisfies the condition above.
  (b) Prove that such a function must be one-to-one.

Problem 16.19.
Let $f : A \rightarrow B$ be a function. Prove that if f is onto, then $\{f^{-1}(\{b\}) : b \in B\}$ partitions the set A.

Problem 16.20.
Suppose that $f : X \rightarrow Y$ is a function, and let A_1 and A_2 be subsets of X.
  (a) If $f(A_1) = f(A_2)$, must $A_1 = A_2$?
  (b) Let f be a bijective function. Show that if $f(A_1) = f(A_2)$, then $A_1 = A_2$. Indicate clearly where you use one-to-one or onto. Did you use both?

Problem 16.21.
Suppose that $f : X \rightarrow Y$ is a function, and let B_1 and B_2 be subsets of Y.
  (a) If $f^{-1}(B_1) = f^{-1}(B_2)$, must $B_1 = B_2$?
  (b) Let f be a bijective function. Show that if $f^{-1}(B_1) = f^{-1}(B_2)$, then $B_1 = B_2$. Indicate clearly where you use one-to-one or onto. Did you use both?

Problem 16.22.
Let X be a nonempty set and let A_1 and A_2 be subsets of X. Recall the notation for characteristic function, χ_A, defined in Problem 13.5.

(a) If $\chi_{A_1} = \chi_{A_2}$, must $A_1 = A_2$?

(b) Show that the product $\chi_{A_1} \cdot \chi_{A_2}$, which is defined pointwise on X by $(\chi_{A_1} \cdot \chi_{A_2})(x) = \chi_{A_1}(x) \cdot \chi_{A_2}(x)$, satisfies $\chi_{A_1} \cdot \chi_{A_2} = \chi_{A_1 \cap A_2}$.

(c) Show that $\chi_{A_1} + \chi_{A_2} - \chi_{A_1 \cap A_2} = \chi_{A_1 \cup A_2}$. (In other words, for each $x \in X$, we have $\chi_{A_1}(x) + \chi_{A_2}(x) - \chi_{A_1 \cap A_2}(x) = \chi_{A_1 \cup A_2}(x)$.)

(d) Can you find a similar result for $\chi_{X \setminus A_1}$?

17

Mathematical Induction

Suppose that you want to show something is true for all positive integers. You could start by checking that the statement is true for $n = 1$, $n = 2$, and so on, but you would have to stop somewhere. Even if you check lots and lots of integers, you can run into problems. Consider the following:

Let us suppose that you are asked to prove that $n^2 + n + 41$ is prime for every positive integer n. You might think the following is good enough to convince someone: if $f(n) = n^2 + n + 41$, then $f(1) = 43$ (which is prime), $f(2) = 47$ (which is prime), $f(3) = 53$ (prime too), and so on. In fact checking the first thirty-nine integers reveals that $f(n)$ is indeed prime for $n = 1, \ldots, 39$. Is this enough evidence to prove that it is true for all positive integers n? Check $n = 40$: $f(40) = 1681$, which is divisible by 41. What's the moral of this story? That examples, even many, many examples, are not a method of proof. It can help us find counterexamples or it can motivate us to formulate a conjecture, but unless we can check every single case, it will never prove anything.

One mathematical technique to prove that a statement holds for all positive integers is to show that the statement is true for $n = 1$ and that whenever it is true for a positive integer n, it is true for the next positive integer $n + 1$. Then, since you have shown it is

207

true for $n = 1$, it must be true for $n = 2$ (because it's always true for a successor). Now that the statement is true for $n = 2$, it has to be true for $n = 3$, because 3 is the integer after 2, and so on. This is called mathematical induction, and a more precise description of this method of proof is given below.

This method is sometimes compared to lining up dominoes and making them fall down (see H. Steinhaus [81]). What has to happen? The first one has to fall, and every time one falls the one after it must fall. Once this happens, all the dominoes do fall down.

Theorem 17.1 (Principle of mathematical induction).
For an integer n, let $P(n)$ denote an assertion. Suppose that
  (i) (The base step) $P(1)$ is true, and
  (ii) (The induction step) for all positive integers n, if $P(n)$ is true, then $P(n + 1)$ is true.
Then $P(n)$ holds for all positive integers n.

The principle of mathematical induction is a direct consequence of the well-ordering principle of $\mathbb{N}$ we came across in Chapter 12. The proof of Theorem 17.1 will be by contradiction: were the induction principle false, then we could construct a nonempty subset of the natural numbers that would not have a minimum—a contradiction to the well-ordering principle. This is the main idea in the proof that follows.

Proof.
Suppose the induction principle were false. Then there would exist an assertion P that would satisfy conditions (*i*) and (*ii*) of the theorem, but $P(n)$ would be false for some $n \in \mathbb{Z}^+$. So let $A = \{k \in \mathbb{Z}^+ : P(k) \text{ is false}\}$. Our supposition implies that A is nonempty. By the well-ordering principle [p. 135], the set A has a minimum. Let m denote this minimum. By condition (*i*), $m \neq 1$. Since $m \in \mathbb{Z}^+$, it follows that $m \geq 2$. Consider the integer $n = m - 1 \geq 1$. Since $n < m$ and m is the minimum of A, we know that $n \notin A$. Thus $P(n)$ is true. By condition (*ii*), $P(n + 1)$ is true too. But $P(n + 1) = P(m)$, so $P(m)$ must also be true, a contradiction. ∎

Students often mistakenly believe condition (*ii*) says that $P(n)$ is true, and ask why we would state it again as a conclusion. Look carefully at condition (*ii*). Note that it is an implication. We are *not* saying that $P(n)$ is true. We *are* saying that *if* $P(n)$ is true, then $P(n+1)$ is true. The antecedent in this implication is called the **induction hypothesis**.

The next example is one that is associated with Carl Friedrich Gauss. As one version of the story goes, when Gauss was 10 years old his teacher, Herr Büttner, asked the students to sum the integers from 1 to 100. Gauss did it almost instantly. It is believed that he did it by the following method.

Write the sum horizontally forwards and backwards as:

$$1 \ + \ \ 2 \ + \ \ 3 \ + \ \cdots \ + \ 99 \ + \ \ 100$$

$$100 \ + \ 99 \ + \ 98 \ + \ \cdots \ + \ \ 2 \ + \ \ \ 1$$

Now add vertically. When you do this, you will get 101 one hundred times; in other words, you get $(101)(100)$. This is twice the sum that you needed, so the answer must be $(101)(100)/2$. There is nothing special about the integer 100. If you try this with a general positive integer n, you will see that $1+2+3+\cdots+n = n(n+1)/2$ for every positive integer n. What a nice formula! Is something like this formula true for the sums of squares of the first n integers? Indeed it is. We'll give it a rigorous proof using mathematical induction.

Example 17.2.
Using mathematical induction, show that

$$1^2 + 2^2 + \cdots + n^2 = \frac{n(n+1)(2n+1)}{6}$$

for every positive integer n.

Proof.
Let $P(n)$ be the assertion that

$$1^2 + 2^2 + \cdots + n^2 = \frac{n(n+1)(2n+1)}{6}.$$

First we check the base step as follows: $P(1)$ is the statement that $1 = (1(1+1)(2+1))/6$, and this is certainly true.

Now we verify the induction step. Suppose that $P(n)$ holds for $n \in \mathbb{Z}^+$. Thus we suppose that for an $n \in \mathbb{Z}^+$ we have

$$1^2 + 2^2 + \cdots + n^2 = \frac{n(n+1)(2n+1)}{6}. \tag{17.1}$$

We wish to show that $P(n+1)$ holds; that is, that

$$1^2 + 2^2 + \cdots + (n+1)^2 = \frac{(n+1)((n+1)+1)(2(n+1)+1)}{2}.$$

We start by grouping the left side of $P(n+1)$ and then simplify as follows:

$$
\begin{aligned}
& 1^2 + 2^2 + \cdots + n^2 + (n+1)^2 \\
&= \left(1^2 + 2^2 + \cdots + n^2\right) + (n+1)^2 \\
&= \frac{n(n+1)(2n+1)}{6} + (n+1)^2 \quad \text{(by our induction hypothesis (17.1))} \\
&= (n+1)\left(\frac{n(2n+1)}{6} + (n+1)\right) \quad \text{(factor out } n+1) \\
&= (n+1)\left(\frac{2n^2 + 7n + 6}{6}\right) \\
&= (n+1)\frac{(n+2)(2n+3)}{6} \\
&= \frac{(n+1)((n+1)+1)(2(n+1)+1)}{6}.
\end{aligned}
$$

By mathematical induction we conclude that the assertion holds for all positive integers. ∎

Induction proofs must contain certain steps. Look at the proof above and see if you can find each of the steps described below.

(1) You should indicate clearly what you are trying to prove. (2) There is always the base step, in which we check the first assertion. (This need not always begin with $n = 1$; it can begin with $n = 3$, $n = 0$, or even at a negative integer! In fact, as long as what you say is true, it can begin at any integer you want it to begin at.) (3) Then we have the induction step, in which we show that for each $n \in \mathbb{Z}$ that is at least as big as the integer used in the base step, if $P(n)$ is true, then $P(n+1)$ is true.

Of course, you still need to write using complete sentences, and you still need to introduce every variable to the reader when the reader meets it (not after the reader has met it for the first time!). Finally, do tell the reader what the base step is ("First we show the assertion holds for $n = 1$"), what the induction step is ("We suppose that $P(n)$ holds for an $n \in \mathbb{Z}^+$; that is $\cdots$ holds"), and what you will prove ("We will show that $P(n + 1)$ holds; that is $\cdots$ holds"). This is as much for your benefit as it is for the reader's. This step shows you where you will begin and where you will have to end. Then show what you said you will show and indicate clearly where you use the induction hypothesis. End your proof with a concluding sentence.

Many statements proved by induction involve sums or products. We remind you of the standard notation for this. In the following, $k \in \mathbb{Z}$ and $a_k \in \mathbb{R}$. The notation for sum is

$$a_1 + a_2 + a_3 + \cdots + a_n = \sum_{k=1}^{n} a_k \,,$$

and the notation for product is

$$a_1 \cdot a_2 \cdot a_3 \cdots \cdots a_n = \prod_{k=1}^{n} a_k \,.$$

This notation often saves space and makes a statement look neater. For instance, the result we proved in Example 17.2 is:

$$\text{For } n \in \mathbb{Z}^+, \text{ we have } \sum_{k=1}^{n} k^2 = \frac{n(n+1)(2n+1)}{6}.$$

If you are ever unsure about what such a statement says, you will almost certainly find it helpful to rewrite the expression the long way.

Exercise 17.3.
Let $x_1, x_2, \ldots, x_n$ be real numbers. Prove that for $n \in \mathbb{Z}^+$, both of the following hold:
(a) $\left| \prod_{k=1}^{n} x_k \right| = \prod_{k=1}^{n} |x_k|$ and
(b) $\left| \sum_{k=1}^{n} x_k \right| \leq \sum_{k=1}^{n} |x_k|$.

The following exercise illustrates how induction can go awry. It's cute, but not very mathematical. A similar example, but a more mathematical one, appears in the problems. See if you can spot the error in that one.

Exercise 17.4.
All people at Bucknell University have the same color hair.

Not a proof.
Let $P(n)$ be the assertion that every group of n people have the same color hair (as each other). Then $P(1)$ is the statement that one person has the same color hair as herself. This is certainly true. So suppose that $P(n)$ is true; that is, when we have n people, they all have the same color hair. We need to show that this implies that $n + 1$ people in a group have the same color hair. So consider a group of $n + 1$ people. If we look at the first n of them (people 1 through n in the group), by the induction hypothesis they all have the same color hair, which we may as well assume is black for right now. So the first n people all have black hair. Now consider the last n people in this group (people 2 through $n + 1$ in the group). Again, by our induction hypothesis, they all have the same color hair. Those who are in both groups are also in the first group, and therefore have black hair. (See Figure 17.1.) Thus, since all people in the second group have the same color hair, everyone has black hair. By mathematical induction we conclude that all people at Bucknell have the same color hair. What a boring campus. ☐

There must be an error! Exactly where is it? ○

Exercise 17.5.
Use induction to prove that for all natural numbers n, the expression $4^n - 1$ is a multiple of 3.

"*Understanding the problem.*" Well, once again, it's probably a good idea to make sure that we know what everything means here. We need to show that $4^n - 1$ is a multiple of 3 for every natural number n. That means we need to show that there exists an integer k such that $4^n - 1 = 3k$.

Group of first n people, all with black hair.

Group of last n people, all with the same color hair.

FIGURE 17.1 They must all have black hair.

"Devising a plan." The outline is presented to you below and the complete solution appears at the end of the chapter.

1. Say clearly what the assertion $P(n)$ is. (Most mathematicians write this out without labeling the assertion with $P(n)$ explicitly.)
2. Check the base step ($n = 0$).
3. Write out the induction step in the principle of mathematical induction clearly. Make sure you replace $P(n)$ by what it says, and replace $P(n + 1)$ by what it says. This will help you figure out what you are supposing (you are supposing $P(n)$) and what you need to end with (you need to end with $P(n + 1)$).
4. Write out the induction hypothesis; that is, write out what you are assuming to be true.
5. Having done all of the above, look at $4^{n+1} - 1$ and show that it is divisible by 3. Indicate clearly where you use the induction hypothesis.
6. State your conclusion clearly.

Solutions to Exercises

Solution to Exercise (17.3).
We leave the proof of the first part to you.

Proof of (b).

We will use the triangle inequality (Theorem 5.5), which has been proven (by you) in Problem 5.9. Our proof will be by induction on n. For $n \in \mathbb{Z}^+$, let $P(n)$ denote the assertion that $|\sum_{k=1}^{n} x_k| \le \sum_{k=1}^{n} |x_k|$.

The validity of the base step, $n = 1$, is clear.

Now assume that $P(n)$ holds for a positive integer n; that is, assume that for a positive integer n, we have $|\sum_{k=1}^{n} x_k| \le \sum_{k=1}^{n} |x_k|$. We must show that $P(n+1)$ holds; in other words, we must show that $|\sum_{k=1}^{n+1} x_k| \le \sum_{k=1}^{n+1} |x_k|$. But

$$\left| \sum_{k=1}^{n+1} x_k \right| = |(x_1 + \cdots + x_n) + x_{n+1}|$$

$$\le |x_1 + \cdots + x_n| + |x_{n+1}| \quad \text{(by the triangle inequality)}$$

$$= \left| \sum_{k=1}^{n} x_k \right| + |x_{n+1}|$$

$$\le \sum_{k=1}^{n} |x_k| + |x_{n+1}| \quad \text{(by the induction hypothesis)}$$

$$= \sum_{k=1}^{n+1} |x_k|,$$

and the result now follows from the principle of mathematical induction. ∎

Solution to Exercise (17.4).

If the base step is for $n = 1$, then the induction step, $P(n)$ implies $P(n+1)$, needs to be valid *for all* $n \ge 1$. We made the following argument: "Those who are in both groups are also in the first group and therefore they have black hair." This argument is not valid if $n = 1$. In that case, the group of the first n people is disjoint from the group of the last n people. However, our argument requires that some person be in both groups. Hence the reasoning falls apart right where it should: If a second person joins a black-haired person there is no guarantee that he or she will also have black hair.

Solution to Exercise (17.5).

Proof.
For $n \in \mathbb{N}$, let $P(n)$ denote the assertion that $4^n - 1$ is a multiple of 3; that is, there is $k \in \mathbb{Z}$ such that $4^n - 1 = 3k$. We will prove this by induction on n.

We check the *base step*. For $n = 0$ the statement becomes $4^0 - 1 = 0$ is divisible by 3. This is obviously true.

Now we check the *induction step*. We assume that for an $n \in \mathbb{N}$, there exists $k \in \mathbb{Z}$ such that $4^n - 1 = 3k$. We need to show that there exists $l \in \mathbb{Z}$ such that $4^{n+1} - 1 = 3l$. Consider the following calculation:

$$4^{n+1} - 1 = 4 \cdot 4^n - 1 = 3 \cdot 4^n + (4^n - 1) = 3 \cdot 4^n + 3k = 3(4^n + k),$$

where the second to last equality is justified by the induction hypothesis. Now set $l = 4^n + k$. Then $l \in \mathbb{Z}$ and $4^{n+1} - 1 = 3l$. Hence the induction step is established.

By the principle of mathematical induction, $4^n - 1$ is divisible by 3 for all $n \in \mathbb{N}$. ∎

Problems

Problem 17.1.
Use the principle of mathematical induction to prove that $1 + 3 + 5 + \cdots + (2n - 1) = n^2$, for every positive integer n.

Problem 17.2.
Use the principle of mathematical induction to prove that $1^3 + 2^3 + \cdots + n^3 = (1 + 2 + \cdots + n)^2$, for every positive integer n.

Problem 17.3.
Use the principle of mathematical induction to prove that $1 + 2 + \cdots + n = n(n + 1)/2$, for every positive integer n.

Problem 17.4.
Prove that if $n \in \mathbb{Z}^+$ and r is a real number such that $r \neq 1$, then

$$\sum_{k=0}^{n-1} r^k = \frac{1 - r^n}{1 - r}.$$

Problem 17.5.
Show that $2^n \leq n!$ for all integers with $n \geq 5$.

Problem 17.6.
Use induction to prove Bernoulli's inequality: For $x \in \mathbb{R}$, if $1 + x > 0$, then $(1 + x)^n \geq 1 + nx$ for all $n \in \mathbb{N}$.

Problem 17.7.
Show that for every positive integer n,

$$\frac{1}{1} + \frac{1}{2} + \frac{1}{3} + \cdots + \frac{1}{2^n} \geq 1 + \frac{n}{2}.$$

(This can be used to show that the harmonic series $1 + \frac{1}{2} + \frac{1}{3} + \cdots + \frac{1}{n} + \cdots$ diverges.)

Problem 17.8.
Show that $2^n > n^2$ for all integers n with $n \geq 5$.

Problem 17.9.
Prove that 8 divides $5^{2n} - 1$ for all $n \in \mathbb{N}$.

Problem 17.10.
Suppose that $g : \mathbb{N} \to \mathbb{N}$ satisfies $g(n+1) = g(n) + g(1)$ for all $n \in \mathbb{N}$.
 (a) Find $g(0)$.
 (b) Show that $g(n + m) = g(n) + g(m)$ for all $n, m \in \mathbb{N}$.

Problem 17.11.
Let $g : \mathbb{N} \to \mathbb{R}^+$ and let a be a positive real number. Suppose that g has the properties that $g(1) = a$ and $g(m + n) = g(m)g(n)$ for all natural numbers n and m.
 (a) Prove that $g(0) = 1$.
 (b) Prove that $g(n) = a^n$ for all $n \in \mathbb{N}$.

Problem 17.12.
Let $a_1, a_2, \ldots, a_n$ be real numbers satisfying $|a_j| \leq 1$ for all $j = 1, 2, \ldots, n$. Prove that for all $n \in \mathbb{Z}^+$ the following holds:

$$\left| \prod_{j=1}^{n} a_j - 1 \right| \leq \sum_{j=1}^{n} |a_j - 1|.$$

Problem 17.13.
Find the error in the *Not a proof* below. (See Problem 10.8 for the definition of the degree of a polynomial.)

Nontheorem.
Let p be a polynomial of positive degree n such that p is a product of degree one polynomials and $p(0) = 0$. If $c \in \mathbb{R}$ satisfies $p(c) = 0$, then $c = 0$.

In other words, our claim is that if $p(x) = ax(a_1x + b_1) \cdots (a_{n-1}x + b_{n-1})$, where $a, b_1, \ldots, b_{n-1} \in \mathbb{R}$ and $a, a_1, \ldots, a_{n-1} \neq 0$, then the only root of p is 0.

Not a proof.
We will prove this statement by induction on the degree n of the polynomial p.

For the base step, we let $n = 1$. Since $p(0) = 0$, we can write $p(x) = ax$ for some $a \in \mathbb{R}$ and $a \neq 0$. If $p(c) = 0$, then $p(c) = ac = 0$. Since $a \neq 0$, we conclude that $c = 0$.

For the induction step, assume that if p is a polynomial of degree n that is a product of degree one polynomials and satisfies $p(0) = 0$, then $p(c) = 0$ implies that $c = 0$. Let p be a polynomial of degree $n + 1$ that factors into $n + 1$ degree one polynomials and satisfies $p(0) = 0$. We need to show that $p(c) = 0$ implies that $c = 0$. Write $p(x) = ax(a_1x + b_1) \cdots (a_nx + b_n)$, where $a, a_1, \ldots, a_n$ are nonzero real numbers and $b_1, \ldots, b_n \in \mathbb{R}$. Suppose that $p(c) = 0$. Then

$$0 = p(c) = ac(a_1c + b_1) \cdots (a_nc + b_n).$$

One of the factors, $ac, a_1c + b_1, \ldots, a_nc + b_n$, must vanish. Rearranging terms if necessary, we may assume that the factor ac or the factor

$a_1c + b_1$ vanishes. Now,

$$q(x) = ax(a_1x + b_1) \cdots (a_{n-1}x + b_{n-1})$$

is a polynomial of degree n that is a product of degree one polynomials and satisfies $q(0) = 0$. Since $ac(a_1c + b_1) = 0$, we have $q(c) = 0$. Since our induction hypothesis applies to q, we conclude that $c = 0$. Therefore, $p(c) = 0$ implies that $c = 0$, and the nontheorem follows from mathematical induction. ▨

There is an equivalent form of the principle of mathematical induction, namely:

Theorem 17.6 (Second principle of mathematical induction).
For an integer n, let $Q(n)$ denote an assertion. Suppose that
  (i) $Q(1)$ is true and
  (ii) for all positive integers n, if $Q(1), \ldots, Q(n)$ are true, then $Q(n+1)$ is true.
Then $Q(n)$ holds for all positive integers n.

Problem 17.14.
Prove the second principle of mathematical induction (Theorem 17.6) from the first one (Theorem 17.1). To do so, let $P(n)$ be the assertion "$Q(1), \ldots, Q(n)$ are true."

Problem 17.15.
Prove that every integer n, where $n \geq 2$, is the product of prime numbers. (We have used this before; this shows that every integer $n \geq 2$ can be factored as a product of primes. If you also proved the uniqueness of this factorization, you have proved the fundamental theorem of arithmetic.)

Problem 17.16.
A subset S of $\mathbb{R}^2$ is **convex** if for every two points $x, y \in S$, the line segment joining x and y again lies in S. Recall that an interior angle at a vertex of a convex polygon is the smaller of the two angles formed by the edges at that vertex.

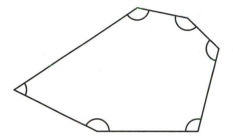

FIGURE 17.2 The sum of all the interior angles in this convex polygon is $4 \cdot 180°$.

Prove that for an integer n, where $n \geq 3$, the sum of all the interior angles of a convex polygon with n vertices is $(n - 2)180$ degrees. (See Figure 17.2.)

Problem 17.17.
Let p_n be a polynomial with real coefficients and of positive degree n. (See Problem 10.8 for the definitions.)
  (a) Suppose $p_n(x) = a_n x^n + a_{n-1} x^{n-1} + \cdots + a_1 x + a_0$. For a real number a, what is the largest the degree of q_n, defined by $q_n(x) = p_n(x) - (x - a)a_n x^{n-1}$, can be?
  (b) Let $a \in \mathbb{R}$ and $n \in \mathbb{Z}^+$. Prove that $p_n(a) = 0$ if and only if $(x - a)$ is a factor of $p_n(x)$.

Problem 17.18.
A **triangular number**, T_n, is the number of equally spaced points that can be used to form an equilateral triangle with sides built of n equally spaced points (see Figure 17.3).
  (a) Find a formula for the nth triangular number, and prove that your formula is correct.
  (b) Can you think of a (familiar) game that uses T_4? T_5?

For $n \in \mathbb{N}$, we define n **factorial**, written as $n!$, as follows. For $n = 0$, $0! = 1$. For $n \geq 0$, define $(n + 1)! = (n + 1) \cdot n!$.

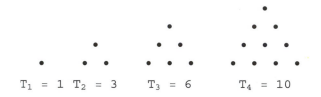

$T_1 = 1$ $T_2 = 3$ $T_3 = 6$ $T_4 = 10$

FIGURE 17.3   Triangular numbers.

For $k, n \in \mathbb{N}$ with $k \leq n$ we define the **binomial coefficient** as

$$\binom{n}{k} = \frac{n!}{k!(n-k)!}.$$

Because the binomial coefficient is the number of ways that we can choose k different elements from a set of n elements, we read $\binom{n}{k}$ as "n choose k."

Theorem 17.7 (Binomial theorem).
Let $a, b \in \mathbb{R}$ and $n \in \mathbb{Z}^+$. Then

$$(a+b)^n = \sum_{k=0}^{n} \binom{n}{k} a^k b^{n-k}.$$

Problem 17.19.
This problem refers to the notation and theorem above.
  (a) Compute each of the following:

$$5!, \quad \binom{8}{3}, \quad \binom{8}{5}, \quad \binom{5}{2}, \quad \binom{5}{3}, \quad \binom{7}{0}, \quad \text{and} \quad \binom{7}{7}.$$

  (b) A "picture proof" of a special case of Theorem 17.7, namely $(m+1)^2 = m^2 + 2m + 1$ for $m \in \mathbb{N}$, is presented in Figure 17.4. Explain the picture proof.
  (c) Prove that for all $k, n \in \mathbb{N}$ with $1 \leq k \leq n$, we get

$$\binom{n+1}{k} = \binom{n}{k-1} + \binom{n}{k}.$$

(If you write out what it means, life will be a lot easier.)

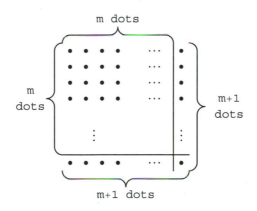

m dots

m dots

m+1 dots

m+1 dots

FIGURE 17.4 "Proof" of $(m + 1)^2 = m^2 + 2m + 1$.

(d) Use part (c) to prove Theorem 17.7. (See Project 27.7, on Pascal's triangle.)
(e) Prove that

$$\sum_{k=0}^{n} \binom{n}{k}(-1)^k = 0 \text{ for all } n \in \mathbb{Z}^+.$$

Problem 17.20.
Deduce the well-ordering principle of $\mathbb{N}$, stated in Chapter 12, from Theorem 17.6 stated on page 218. (Recall that we used the well-ordering principle to prove Theorem 17.1. Since Theorem 17.6 and Theorem 17.1 are equivalent, this problem shows that the principle of mathematical induction and the well-ordering principle of $\mathbb{N}$ are equivalent.)

18

Sequences

CHAPTER

We have all seen lists of numbers. For example, we've all worked with a list of positive even integers presented in increasing order, $(2, 4, 6, \ldots, 2n, \ldots)$, where $n = 1, 2, 3, \ldots$. The positive odd numbers $(1, 3, 5, \ldots, 2n - 1, \ldots)$ can also be presented in such a list, where $n = 1, 2, 3, \ldots$. What we are interested in here is a precise definition of "infinite list."

Here's an example from your childhood of a problem that yields such a list. Let n be a positive integer with $n \geq 2$. Suppose that we have n children arranged in a circle, and that rather than use their names, we number them $\{1, 2, \ldots, n\}$. Say these children want to see who goes first in a game. They begin by eliminating the second child, and then proceed around the circle, eliminating every other child until there is only one child left. That happy child goes first. The question is, where should you stand in order to be the winning child? Let's start small: if there are two children, you should stand in the first spot. If there are three, you should stand in the third spot. If there are four, you should stand in the first spot. Where should you stand if there are n children? (This challenging problem is known as the Josephus problem. The answer appears at the end of the chapter. Of course, the children can count off by three or four, giving us a new problem to solve.) In this problem, for each group of n children,

223

we have an answer. Thus we again have a list of numbers. We now turn to the definition of "list."

A **sequence** is a function f from the natural numbers $\mathbb{N}$ to a set X. In this chapter, we will concentrate on the case $X = \mathbb{R}$. It is standard to write $x_n = f(n)$, and to refer to x_n as a **term** in the sequence. The sequence will be denoted $(x_n)_{n=0}^{\infty}$, or just (x_n) when it is clear which n we are referring to or when it doesn't matter where the sequence starts. We can begin a sequence at an integer other than $n = 0$ when convenient, and we will often begin the sequence at $n = 1$ without much fanfare. We'll tell you where we are starting when it really matters.

Since sequences are functions, we can graph them as functions defined on the nonnegative (or positive) integers and then we can see what they are doing.

Example 18.1.
The sequence (x_n) is defined by $x_n = 1 + 1/n$ for $n \in \mathbb{Z}^+$. We will write out the first few terms and graph the beginning of the sequence in Figure 18.1.

The first four terms are: $x_1 = 2$, $x_2 = \frac{3}{2}$, $x_3 = \frac{4}{3}$, $x_4 = \frac{5}{4}$, ...○

Exercise 18.2.
For each of the sequences given below, write out the first four terms and graph each as a function from its domain (a subset of $\mathbb{Z}$) to its codomain.

(a) Let $x_n = 1 - (-1)^n$, for $n = 0, 1, 2, \ldots$.

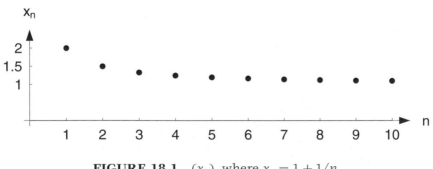

FIGURE 18.1 (x_n), where $x_n = 1 + 1/n$.

(b) Let $x_n = n/(n+1)$, for $n = 0, 1, 2, \ldots$.
(c) Let $x_n = (n^2 + 1)/(1 - n)$, for $n = 2, 3, \ldots$. ○

When we look at sequences, we notice that different sequences behave in different ways. This is illustrated by our examples in Exercise 18.2. Some sequences approach some horizontal line as $n \to \infty$. Some seem to be shooting off to infinity, others jump around a lot. We need to understand how these sequences differ from each other, and the following definitions will help us do that.

You'll notice that a lot of terms that we introduced when we studied sets reappear here. What is really happening is that each sequence (x_n) (where order counts) gives rise to a nonempty set $S = \{x_n : n \in \mathbb{N}\}$ (where order doesn't count). For example, the sequence (z_n) where $z_n = (-1)^n$ gives rise to the set $S = \{-1, 1\}$, while the sequence $(y_n)_{n=1}^\infty$ where $y_n = n$ gives rise to the set $T = \{n : n \in \mathbb{Z}^+\} = \mathbb{Z}^+$. So we can ask when a sequence is bounded above, bounded below, bounded, has an infimum or supremum—and we can use everything we learned about sets to find an answer.

For example, a sequence of real numbers (x_n) is bounded if the set $S = \{x_n : n \in \mathbb{N}\}$ is bounded. Thus, according to our definition of bounded set, a sequence of real numbers (x_n) is **bounded above**, if there is a real number M such that $x_n \leq M$ for all n, and **bounded below**, if there exists a real number m such that $x_n \geq m$ for all n. A sequence is **bounded** if it is bounded above and below. That's how we defined it. But you may find that the following provides a more useful way to think about boundedness in $\mathbb{R}$.

Exercise 18.3.
Let (x_n) be a sequence. Prove that (x_n) is bounded if and only if there exists a real number N such that $|x_n| \leq N$ for all n.

Just as before, a real number U satisfying $x_n \leq U$ for all n is called an **upper bound** of the sequence (x_n), and a real number L satisfying $L \leq x_n$ for all n is a **lower bound** of the sequence (x_n). In our illustration above, we considered the sequence (z_n), where $z_n = (-1)^n$. We see that this sequence is bounded above (1 is an upper bound) and bounded below (-1 is a lower bound), and therefore it is bounded. Alternatively, $|z_n| \leq 1$ for all n, and by the previous exercise we may conclude that the sequence (z_n) is bounded.

Exercise 18.4.
  (a) Give an example of a real number that is not an upper bound of the sequence given by $x_n = n/(n+1)$.
  (b) Complete the following sentence: The real number U is not an upper bound of the sequence (x_n), if ....   ○

    If (x_n) is a sequence that is bounded below, the set $S = \{x_n : n \in \mathbb{N}\}$ is bounded below. Since S is a nonempty set of real numbers that is bounded below, the infimum version of the completeness axiom (see Exercise 12.8) implies that S has an infimum, and we call this the **infimum** of the sequence (x_n) (or **greatest lower bound** of (x_n)). We denote it by $\inf(x_n)$. Recall that you showed (Problem 12.14) that the infimum is unique.

Example 18.5.
Let (x_n) be a sequence that is bounded below. State the two properties that the infimum of (x_n) must satisfy.

    The infimum of the sequence (x_n) is the real number m satisfying

  (i) $m \leq x_n$ for all n, and
  (ii) if p is a real number satisfying $p \leq x_n$ for all n, then $p \leq m$.

    Each of the statements (i) and (ii) can be stated in the vernacular, and you should do so now.   ○

Exercise 18.6.
Guess the infimum for each of the cases below:
  (a) $x_n = 1/n$, for $n = 1, 2, 3, \ldots$;
  (b) $x_n = n^2$, for $n \in \mathbb{N}$;
  (c) $x_n = n/(n+1)$, for $n \in \mathbb{N}$;
  (d) $x_n = (-1)^n/(n^2+1)$, for $n \in \mathbb{N}$.   ○

    When you look for the infimum, remember that it may or may not appear in the sequence. Everything we do for the infimum can be done for the supremum. So the **supremum** of the sequence (x_n) (or **least upper bound** of (x_n)) is the supremum of the set $S = \{x_n : n \in \mathbb{N}\}$. We denote it by $\sup(x_n)$. The rest is left to you in the next exercise.

Exercise 18.7.
Let (x_n) be a sequence that is bounded above. State the two properties that the supremum of (x_n) must satisfy, and say why it exists. ○

That takes care of how high and how low a sequence can go. Now we turn to how it gets where it is going.

A sequence (x_n) is **increasing** if $x_n \leq x_{n+1}$ for all n, and **decreasing** if $x_n \geq x_{n+1}$ for all n. We say the sequence (x_n) is **strictly increasing** if $x_n < x_{n+1}$ for all n. Likewise, a sequence (x_n) is **strictly decreasing** if $x_n > x_{n+1}$ for all n.

Exercise 18.8.
The object of this exercise is to make sure you understand the definitions above. Either explain why you cannot give an example of the following, or give an example of
  (a) a bounded sequence. Find an upper bound and a lower bound.
  (b) a sequence that is bounded below, but not bounded above. Find a lower bound. Must the sequence be increasing?
  (c) a sequence that is bounded above, but not bounded below. Find an upper bound.
  (d) an increasing sequence that is neither bounded above nor below.
  (e) a strictly increasing bounded sequence.
  (f) a strictly decreasing sequence that is bounded above, but not below.
  (g) a sequence that is neither strictly increasing nor strictly decreasing. ○

Exercise 18.9.
What is your best guess for the supremum of the sequence $x_n = \underbrace{0.999\ldots9}_{n\ 9's}$? In the next chapter we will carefully determine the supremum. ○

Since sequences are functions, we can manipulate them algebraically. For example, to add two sequences together, we define the sum $(x_n) + (y_n)$ to be the sequence $(x_n + y_n)$. In the same way, we may subtract sequences.

Example 18.10.
Suppose (x_n) and (y_n) are two bounded sequences. If $\sup(x_n) = l$ and $\sup(y_n) = m$, is $\sup(x_n + y_n) = l + m$?

We'll first show that $l + m$ is an upper bound of $(x_n + y_n)$. (Thus it makes sense to talk about the supremum of $(x_n + y_n)$.) Then we will try to show the second thing: that $l + m$ is the least of all the upper bounds, in the sense defined above. Remember that since we don't know the answer here, our attempt might fail.
So let $l = \sup(x_n)$ and $m = \sup(y_n)$. Then $x_n \le l$ for all n and $y_n \le m$ for all n. Thus $x_n + y_n \le l + m$ for all n. So far so good; we know that $l + m$ is an upper bound (and we know that the sequence $(x_n + y_n)$ is bounded above). But we still have to see whether or not $l + m$ is the least upper bound. So suppose that p is another upper bound. We are supposed to show that $p \ge l + m$. Well, $p \ge x_n + y_n$ for every n, but that doesn't seem to help since that doesn't (in general) imply anything about the relation between p and x_n or p and y_n. In fact, closer inspection reveals that if one of the sequences is negative, we can't say anything at all. So we abandon our attempt at a proof and search for a counterexample, using what we learned above.
Let (x_n) be a bounded nonconstant sequence, say $x_1 = 1$ and $x_n = 2$ for all other n. Thus $l = 2$. Now let $y_n = -x_n$. Then $m = -1$. So $x_n + y_n = 0$, and the supremum of $(x_n + y_n)$ is clearly 0, while $l + m = 1$. Hence the supremum of $(x_n + y_n)$ need not be $l + m$. ○

The lesson here is that in trying to prove something, we came up with an example that showed it wasn't true. This is a perfectly reasonable way to approach the problem as long as we are always on the lookout for what can go wrong with a proof.

Exercise 18.11.
Let (x_n) be a sequence that is bounded below. Let $l = \inf(x_n)$. Show that $(-x_n)$ is bounded above and find the supremum of the sequence. ○

Functions defined on $\mathbb{N}$ (or sequences) are sometimes defined by a process called recursion, which is mathematical induction's way of defining such a function. To put it as simply as we can, **recursion** is done by first deciding where you want to start the sequence (we'll call this the initial integer). We then specify the value of the sequence at the initial integer to obtain the first term. Finally, we write out a rule for computing the $(n+1)^{st}$ term, $f(n+1)$, given the nth term, $f(n)$, for each integer n at least as big as the initial integer.

The most familiar example is probably the factorial function. Recall that we define the factorial function, $f(n) = n!$, by $f(0) = 1$, and by giving the rule for finding $f(n+1)$: $f(n+1) = (n+1)f(n)$ for all $n \in \mathbb{N}$. Now given a natural number, you should be able to find $n!$ by computing $f(n)$. To understand such a function, it's always a good idea to write out the first few terms.

Exercise 18.12.
Suppose that $f(0) = 1$ and $f(n+1) = 2f(n)$ for every $n \in \mathbb{N}$. Then f is a familiar function. What is it? ○

One of the most famous examples of a sequence defined recursively is the Fibonacci sequence. Fibonacci, whose real name is Leonardo Pisano, was born in 1170 in Pisa. (One source you might consult for more information about Leonardo Pisano is L. Sigler's book [77].) The Fibonacci sequence is often presented with pictures of rabbits. So here is a version of Fibonacci's original rabbit problem: Suppose that rabbits live forever. Starting at the age of two months, each pair produces (exactly) one baby pair, and continues to do so every month thereafter. If we start with one brand new pair, how many pairs of rabbits will we have in the nth month? Now here's the sequence. See if you can figure out the reference to these rabbits.

Define $F_0 = 0, F_1 = 1$, and $F_n = F_{n-1} + F_{n-2}$ for $n \geq 2$. This sequence is called the **Fibonacci sequence** and the terms of the sequence are called the **Fibonacci numbers**.

Example 18.13.
Starting with $n = 1$, find the first 6 terms of the Fibonacci sequence.

The first six Fibonacci numbers are: $F_1 = 1$, $F_2 = 1$, $F_3 = 2$, $F_4 = 3$, $F_5 = 5$, $F_6 = 8$. ○

The Fibonacci sequence is extremely appealing to mathematicians and nonmathematicians. In fact, there are many web sites and books with information and problems about Fibonacci sequences, as well as the journal *The Fibonacci Quarterly*.

We present one of the many interesting patterns found in Fibonacci numbers below. Others can be found in the problems, as well as some of the references given in this chapter.

Exercise 18.14.
Let (F_n) denote the Fibonacci sequence. Show that $F_{n+1}F_{n-1} - F_n^2 = (-1)^n$ for every positive integer n.

Check the equation for the first few values of n to see if this is reasonable (but, of course, this is not a proof). The proof of Exercise 18.14 will use mathematical induction, and you can read our solution below when you are ready. ○

We now return to the solution of the Josephus problem mentioned at the beginning of the chapter. For each integer $n \geq 2$, we let $f(n)$ denote the number of the winning child. Then $f(2) = 1$, $f(3) = 3$, $f(2n) = 2f(n) - 1$, and $f(2n + 1) = 2f(n) + 1$. For more information on the Josephus problem, we recommend the article [76].

Solutions to Exercises

Solution to Exercise (18.3).
If you have solved Problem 12.7, then you have solved this exercise as well. If not, here is a solution.

First we'll prove that if (x_n) is bounded, then there exists a real number N such that $|x_n| \leq N$ for all n. By the definition of bounded sequence, there exist real numbers m and M such that $m \leq x_n \leq M$, for all n. Hence $-|m| \leq m \leq x_n \leq M \leq |M|$ for all n. Letting $N = \max\{|m|, |M|\}$, we have $-N \leq x_n \leq N$. Thus $|x_n| \leq N$ for all n.

We'll leave the proof that "(x_n) is bounded if there exists an integer N with $|x_n| \leq N$ for all n" to you.

Solution to Exercise (18.4).
Many answers are possible for (a).
  (a) Consider the number $m = -1$. Then m is not an upper bound of the sequence since $x_1 > m$.
  (b) A real number U is not an upper bound of the sequence $(x_n)_{n \in \mathbb{N}}$ if there exists $n \in \mathbb{N}$ such that $x_n > U$.

Solution to Exercise (18.6).
The answers are: (a) 0, (b) 0, (c) 0, (d) $-1/2$.

Solution to Exercise (18.7).
A real number U is the supremum of a sequence (x_n) if (i) $x_n \leq U$ for all n, and (ii) if V is another upper bound of (x_n), then $U \leq V$.

Solution to Exercise (18.8).
You should be able to find examples for all parts of this problem, except part (d). An increasing sequence will always be bounded below, and its first term will serve as a lower bound. For parts (a) and (g), the sequence $((-1)^n)$ yields such an example. For (c) and (f), you can use the sequence $(-n)$, which is bounded above by 0 but is not bounded below. For (b), the sequence defined by $x_n = n + 2(-1)^n$ for $n \in \mathbb{N}$ is bounded below (by, for example, -100) and this sequence is not increasing (since $x_0 = 2$ and $x_1 = -1$). Finally, for (e) the sequence $(1 - 1/n)$ for $n \in \mathbb{Z}^+$ serves as an example.

Solution to Exercise (18.11).
Since $l = \inf(x_n)$, we know that $x_n \geq l$ for all n. Multiplying both sides by -1 we obtain $-x_n \leq -l$ for all n, and consequently $(-x_n)$ is bounded above and $-l$ is an upper bound. We claim that $-l$ is the supremum, too. Suppose that m is also an upper bound. Then $-x_n \leq m$ for all n. Multiplying by -1, we see that $x_n \geq -m$ for all n. But this implies that $-m$ is a lower bound for (x_n). Since $l = \inf(x_n)$, we know that $-m \leq l$. Thus $m \geq -l$, and $-l$ is the least of all the upper bounds. So $-l = \sup(-x_n)$.

Solution to Exercise (18.12).
We note that $f(0) = 1, f(1) = 2, f(2) = 4$ and $f(3) = 8$. We guess that $f(n) = 2^n$ for all $n \in \mathbb{N}$. This can (and should) be rigorously proved using induction.

Solution to Exercise (18.14).
For $n = 1$, we easily check that $F_2 F_0 - F_1^2 = -1$. Similarly, for $n = 2$, we can check that $F_3 F_1 - F_2^2 = 1$. Now suppose that for an arbitrary $n \in \mathbb{N}$ with $n \geq 2$, we have $F_{n+1} F_{n-1} - F_n^2 = (-1)^n$. In other words, we assume that $F_n^2 = F_{n+1} F_{n-1} - (-1)^n$. We will show that $F_{n+2} F_n - F_{n+1}^2 = (-1)^{n+1}$. To see this, note that $F_{n+2} = F_{n+1} + F_n$ for all n. Thus (you should fill in reasons for each of the equalities):

$$F_{n+2} F_n - F_{n+1}^2 = (F_{n+1} + F_n) F_n - F_{n+1}^2$$
$$= F_{n+1} F_n + F_n^2 - F_{n+1}^2.$$

Use the induction hypothesis to replace the middle term, F_n^2, in the summand above to conclude that

$$F_{n+2} F_n - F_{n+1}^2 = F_{n+1} F_n + F_{n+1} F_{n-1} - (-1)^n - F_{n+1}^2$$
$$= F_{n+1}(F_n + F_{n-1}) - F_{n+1}^2 + (-1)^{n+1}$$
$$= (-1)^{n+1},$$

and the result now follows from the principle of mathematical induction. ∎

The first 100 Fibonacci numbers can be found on the web [49]. From there you can get to a very cute picture of little rabbits, as well as a set of puzzles based on these numbers. Fibonacci numbers also make an appearance in the popular children's book *The Number Devil* by Hans Magnus Enzensberger [21].

Problems

Problem 18.1.
Graph the following sequences and briefly describe each of the graphs:
  (a) $x_n = (-1)^n$, where $n \in \mathbb{N}$;
  (b) $x_n = 1/2^n$, where $n \in \mathbb{N}$;
  (c) $x_n = n/(n-1)$, where $n \in \mathbb{N}$ and $n \geq 2$;
  (d) $x_n = (-1)^n/2^n$, where $n \in \mathbb{N}$;
  (e) $x_n = (-1)^n(n^2/(n+1))$, where $n \in \mathbb{N}$.

Problem 18.2.
  (a) Find an example of each of the following:
      (i) a strictly increasing sequence;
      (ii) an increasing sequence that is not strictly increasing;
      (iii) a bounded strictly increasing sequence;
      (iv) a sequence that is not increasing.
  (b) Now find four interesting examples that correspond to the ones in part (a) but replace increasing by decreasing.

Problem 18.3.
  (a) Give an example of a sequence of rational numbers that is bounded above.
  (b) Give an example of a sequence of rational numbers that has no upper bound, but does have a lower bound.
  (c) Give an example of a strictly increasing sequence of numbers that has a supremum, but such that the supremum is not a term in the sequence. Can you find a strictly increasing sequence such that the supremum is equal to x_n for some n? Why or why not?

Problem 18.4.
Give an example of a sequence of rational numbers that has an irrational number as supremum.

Problem 18.5.
If $l = \sup(x_n)$, what is $\inf(-x_n)$? (You should know by now that the first thing to do is to try examples. Make up at least three different examples.) State your conjecture. Prove it.

Problem 18.6.
If $l = \sup(x_n)$, what is $\sup(k x_n)$ where $k \in \mathbb{R}^+$? Prove your conjecture.

Problem 18.7.
Let (x_n) be a bounded sequence such that $x_n \leq -2$ for all $n \in \mathbb{N}$.
  (a) Prove that (x_n^2) is bounded.
  (b) Let $l = \inf(x_n)$ and $m = \sup(x_n)$. Find $\inf(x_n^2)$ in terms of l or m, or both. Prove that your result is correct.

Problem 18.8.
Suppose that (x_n) and (y_n) are bounded below.
  (a) Show that $\inf(x_n + y_n) \geq \inf(x_n) + \inf(y_n)$.
  (b) Is it always true that $\inf(x_n + y_n) = \inf(x_n) + \inf(y_n)$? Prove this or give a counterexample.

Problem 18.9.
Suppose that (x_n) and (y_n) are bounded below. Is it always true that $\inf(x_n y_n) = \inf(x_n) \inf(y_n)$? Prove this or give a counterexample.

Problem 18.10.
Let $f(0) = 2$, $f(1) = 2$ and define $f(n + 1) = f(n)f(n - 1)$. Find a nonrecursive formula for f.

Problem 18.11.
Define a sequence (x_n) by $x_0 = 1000$ and for $n \geq 1$, define $x_n = (.05)x_{n-1}$. Find another representation for this sequence. Have you seen this anywhere else before? If so, where?

Problem 18.12.

Let (F_n) be the Fibonacci sequence and $x_n = F_{n+1}/F_n$, for $n \geq 1$. Show that $x_n = 1 + 1/x_{n-1}$, for $n \geq 2$.

Problem 18.13.

Let (F_n) denote the Fibonacci sequence. Define the Lucas sequence by $L_0 = 2$, $L_1 = 1$, and for $n \geq 2$ define $L_{n+1} = L_n + L_{n-1}$. (For some proofs you may want to use the second principle of induction stated in the problem section of Chapter 17 as Theorem 17.6.)

(a) Calculate $L_1, \ldots, L_{10}$.
(b) Calculate $L_n - F_{n-1}$, for $n \geq 1$. Find a remarkable pattern in this list of numbers. State it clearly and prove it by induction.
(c) Calculate $F_n + L_n$. Find a remarkable pattern in this list of numbers. State it clearly and prove it using part (b).

19

CHAPTER

Convergence of Sequences of Real Numbers

As we saw in the last chapter, when we graph several terms of a sequence, certain behavior may appear. We may become convinced, for whatever reason, that the sequence is unbounded. Or, we may believe that the sequence is bounded and we may even notice the sequence moving towards a particular horizontal line. But how do we check that what we believe is happening really is happening?

Our efforts to explain this require that you fully understand how to measure distance. So we remind you that distance is usually measured using the absolute value function, or $|x|$, and the absolute value of a real number x measures the distance from x to 0. If we want to measure the distance between two real numbers x and a, we would need to look at $|x - a|$.

For an arbitrary positive real number ϵ, we know what it means to say $|x| = \epsilon$. What does it mean to say $|x| < \epsilon$? The answer is, as you can check, that $|x| < \epsilon$ if and only if $-\epsilon < x < \epsilon$. So how do we determine when a sequence of real numbers approaches a real number L? We will use the absolute value function to measure the distance from terms in the sequence to L. We make this precise in the following definition.

We say that a sequence (x_n) **converges** if there exists a real number L such that for all $\epsilon > 0$ there exists a real number N such that $|x_n - L| < \epsilon$ for all $n \geq N$. If such an L exists, we call L the **limit** of the sequence (x_n), we say that (x_n) **converges** to L, and we write $x_n \to L$ or $\lim_{n\to\infty} x_n = L$. If no such L exists, we say that the sequence **diverges**. While we allow N to be a real number, we caution you to remember that the indices on the terms of a sequence, x_n, are natural numbers. To really understand this definition, we must understand it visually, and we must also know how to use it to show that a sequence converges. We first turn to the visual aspect of convergence.

Let's think about the definition. If we believe the sequence (x_n) converges, then we need to find a real number L such that the sequence gets really really close to the line $y = L$. Mathematically, we say that we need the sequence arbitrarily close to the horizontal line. This, in turn, implies that the distance from the sequence to the line $y = L$ should be less than every positive real number ϵ that we can think of; that is what arbitrarily close means. But it may not happen right away; it may only happen eventually (for each ϵ, there will exist N such that x_n may not satisfy what we want for $n < N$, but it will satisfy it for $n \geq N$). That's why we defined convergence the way we did.

We illustrate this definition in Figure 19.1. In this picture, we first pick a value $\epsilon = \epsilon_1 > 0$. Then we indicate a corresponding real number $N = N_1$ such that for all $n \geq N_1$ we have $|x_n - L| < \epsilon_1$.

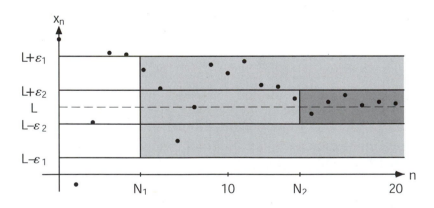

FIGURE 19.1 Definition of a convergent sequence; two-dimensional illustration.

FIGURE 19.2 Definition of a convergent sequence; one-dimensional illustration.

Looking at the figure, we see that the strip from $y = L - \epsilon_1$ to $y = L + \epsilon_1$ contains all the terms of the sequence from x_5 on. Generally, the smaller the value of ϵ, the larger the value of N. Let's think about why this is true: Returning to Figure 19.1, we see that $\epsilon = \epsilon_2$ is smaller than ϵ_1, and this, in turn, forces the sequence to be closer to the horizontal line. So we go farther out in the sequence to get closer to the line $y = L$. You also probably noticed that once we find a value of N that works, anything larger than N will work, too.

Figure 19.2 is yet another way to illustrate the same situation. Explain this sketch to yourself.

Now let's turn to how we show a sequence converges. First we make a conjecture as to the value of the limit, call it L. The important thing to notice, when we make our conjecture, is that we are interested in the behavior as $n \to \infty$. We don't really care what happens for small n, for example. So if we ask what a sequence converges to, you can guess by ignoring terms that don't really matter in the long run. For example, if we ask you to guess what

$$\left(\frac{2n^2 + 3n + 4}{4n^2 - 3n - 5} \right)$$

converges to, you would remember that only behavior near infinity matters, so you would probably guess that $2n^2$ really dominates the quantity in the numerator, while $4n^2$ really dominates the quantity in the denominator. Thus, when you guess this limit, you'll probably guess that it's the same as the limit of the sequence with terms $x_n = 2n^2/(4n^2)$—which it is. You probably also would guess that this limit is $1/2$, and you'd be right again. Before we move to our second step, try practicing your guessing on some of the examples below.

Exercise 19.1.
Guess the limits of each of the following sequences:

(a) $\left(\dfrac{3n^2 + n + 4}{3 + n + 5n^2}\right)$;

(b) $\left(\dfrac{2n^3 + n^2 + 5}{n^4 + 6}\right)$;

(c) $\left(\dfrac{\sqrt{n} + n}{\sqrt{n} - n}\right)$.

Now, once you have guessed your limit, you must prove that it is correct. To do this, assume that we are given an arbitrarily small number, which we denote by ϵ. We then try to find a real number N (depending on ϵ) so that for the remaining terms of the sequence, that is, for $n \geq N$, we have $|x_n - L| < \epsilon$. This is where things get tricky.

Let's see how we would use this definition, starting with a fairly simple example.

Example 19.2.
Show that the sequence (x_n) defined by $x_n = 1/n$ converges to 0.

We'll find this easier if we follow Pólya's list.

"*Understanding the problem.*" We need to show that for every $\epsilon > 0$ there exists a real number N such that $n \geq N$ implies that $|x_n - 0| < \epsilon$. We fill in what we can, remembering that ϵ is arbitrary; that is, it is chosen for us and we have no control over its value. So once someone gives us ϵ, we are supposed to come up with N so that $|1/n - 0| < \epsilon$ for $n \geq N$.

"*Devising a plan.*" Now we'll work backwards to see what N has to be. We need $|1/n| < \epsilon$. So we need $n > 1/\epsilon$. It appears that if we take $N > 1/\epsilon$, say $N = 2/\epsilon$, we'll get exactly what we need when $n \geq N$. Since N depends on ϵ (there's an ϵ in our definition of N), many authors write $N = N(\epsilon)$.

Before we carry out the plan, note that if we knew ϵ, then this would tell us exactly how big N has to be. If $\epsilon = 0.1$, then N needs to be bigger than $1/0.1 = 10$; if $\epsilon = 0.01$, then N needs to be bigger than $1/0.01 = 100$; and if ϵ just happens to be itself, then N needs to be greater than $1/\epsilon$. We're now ready to carry out the plan. We have to write this up, so that a reader who has not seen our work will know what we are doing.

Proof.
If $\epsilon > 0$ and we choose $N = 2/\epsilon$, then for $n \geq N$, we have $|x_n - 0| = |1/n - 0| = 1/n \leq 1/N = \epsilon/2 < \epsilon$. Therefore, if $n \geq N$, then $|x_n - 0| < \epsilon$ as desired. ∎

Our second example is a bit more challenging, and requires slightly different techniques. You'll see more problems of this type and more challenging limit problems when you take your first analysis course.

Example 19.3.
Show that $\lim_{n \to \infty} n/(n+2) = 1$.

As above, we'll first understand our problem, and then devise a plan.

"*Understanding the problem.*" We start by writing out the definition: For all $\epsilon > 0$ there exists N such that $n \geq N$ implies that $|x_n - L| < \epsilon$.

Next we'll fill in x_n and L: For all $\epsilon > 0$ there exists N such that $n \geq N$ implies that $|n/(n+2) - 1| < \epsilon$.

Now simplify the expression $|n/(n+2) - 1| < \epsilon$ to find out how big n must be in terms of ϵ: We need to find N so that $n \geq N$ implies that $2/(n+2) < \epsilon$.

"*Devising a plan.*" You can solve for n, as above, if you wish, but that method only works well in simple cases. So we are going to try to change this problem from the one we have to a simpler problem. How will we do that? Well, we want to make $2/(n+2)$ small. If we can find something bigger and simpler than $2/(n+2)$, and if we can make that less than ϵ, then we will also know that $2/(n+2)$ is less than ϵ. What's bigger and simpler? The previous example wasn't too bad. So if we just had a simple fraction, we would be in good shape. A general strategy that often works is this: If the numerator is complicated, we'll try to find something simpler and larger than the numerator. If the denominator is complicated, we'll try to find something simpler and smaller than the denominator. Our simpler expression must still "act the same in the long run" as the original.

For this exercise, the numerator is simple, so we'll leave it alone. For the denominator, we'll try to somehow use the thing that dominates: n. We need $n + 2$ greater than or equal to something simple involving n. It's pretty clear that $n + 2 \geq n$, so we'll use n. Putting this together, we have found that

$$\frac{2}{n+2} \leq \frac{2}{n}.$$

Therefore, if we make $2/n < \epsilon$, we will also make $2/(n+2) < \epsilon$. But making $2/n < \epsilon$ is easy, since $2/n < \epsilon$ if and only if $n > 2/\epsilon$. Thus, it appears that if $N > 2/\epsilon$, then $2/n < \epsilon$ and therefore $2/(n+2) < \epsilon$, which is what we need.

"*Carrying out the plan.*" Write out the proof, beginning with "If $\epsilon > 0$ and ...." The very next phrase should identify N, unless there are things you need to tell the reader in order for the reader to understand your definition of the real number N. Remember that the reader will only see your proof, not your plan.

Proof.
If $\epsilon > 0$ and we choose $N = 3/\epsilon$, then for $n \geq N$, we have

$$\left| \frac{n}{n+2} - 1 \right| = \left| \frac{2}{n+2} \right| \leq \frac{2}{n} \leq \frac{2}{N} = \frac{2\epsilon}{3} < \epsilon.$$

Thus, $\lim_{n \to \infty} n/(n+2) = 1$. ∎

Here's one more exercise on limits.

Exercise 19.4.
The sequence $\big((2n + 4)/(n^2 + n + 1)\big)$ converges. Guess the limit and prove that your guess is correct, using the definition of convergence. ○

It's time to think about negating the definition of convergence.

Exercise 19.5.
By negating the definition of convergence, explicitly state what it means for a sequence (x_n) to diverge. ○

Exercise 19.6.

Using Exercise 19.5, show that the sequence $((-1)^n)$ diverges. ○

We state the most basic properties of limits and convergent sequences here. The first theorem says that there can be one and only one choice for the limit of a convergent sequence.

Theorem 19.7.

If a sequence converges, then the limit is unique.

We've done uniqueness proofs before and we'll do this one the same way: We suppose to the contrary that there are two different limits L and M, and then we will show that they must be the same. So, we need to show that $L - M = 0$. We also use a standard trick: *we add and subtract the same quantity to an object.* Why? Well, since all we know is that L and M get close to the terms of the sequence (x_n), we have to somehow use these terms. But there is no x_n in the equation $L - M = 0$. So we will have to insert an x_n where none appears.

Proof.

Let (x_n) be a convergent sequence. Suppose to the contrary that $x_n \to L$ and $x_n \to M$, where $L \neq M$. Let $\epsilon = (1/4)|L - M|$. Then $\epsilon > 0$. By the definition of convergence, since $\epsilon > 0$, there exists N_1 such that $|x_n - L| < \epsilon$ for $n \geq N_1$ and there exists N_2 such that $|x_n - M| < \epsilon$ for $n \geq N_2$. Let $N = \max\{N_1, N_2\}$. Then for $n \geq N$ we have

$$
\begin{aligned}
|L - M| &= |L - x_n + x_n - M| \\
&\leq |L - x_n| + |x_n - M| && \text{(by the triangle inequality)} \\
&< \epsilon + \epsilon && \text{(since } n \geq N_1 \text{ and } n \geq N_2\text{)} \\
&= \tfrac{1}{2}|L - M|.
\end{aligned}
$$

But this is silly, since no positive real number is smaller than half of itself. This contradiction establishes the result that limits of sequences are unique. ∎

Here's another important theorem. It uses Exercise 18.3, which says that a sequence (x_n) is bounded if and only if there exists a real number M such that $|x_n| \leq M$ for all n.

Theorem 19.8.
Every convergent sequence is bounded.

Proof.
Suppose that the sequence $(x_n)_{n=1}^\infty$ converges to the real number L. Let $\epsilon = 1$. Then there exists N such that $|x_n - L| < 1$ for all $n \geq N$. Let K be the smallest integer satisfying $K \geq N$. Thus $|x_n| = |x_n - L + L| \leq |x_n - L| + |L| < 1 + |L|$ for all $n \geq K$. Consider the numbers $|x_1|, |x_2|, \ldots, |x_{K-1}|$ and $1 + |L|$. Since there are finitely many such numbers, we may choose the maximum of these. Let $M = \max\{|x_1|, |x_2|, \ldots, |x_{K-1}|, 1 + |L|\}$. Then $|x_n| \leq M$ for all n, and we conclude that the sequence (x_n) is bounded. ∎

Part (i) of the next theorem says that if we sum two convergent sequences, the new sequence converges, too. It also says "the limit of the sum is the sum of the limits." What do the other parts say?

Theorem 19.9.
Let (x_n) and (y_n) be two sequences that converge. Let L and M be real numbers such that $x_n \to L$ and $y_n \to M$. Then
   (i) $x_n + y_n \to L + M$,
   (ii) $\alpha x_n \to \alpha L$, for every real number α,
   (iii) $x_n y_n \to LM$, and
   (iv) if $M \neq 0$ and $y_n \neq 0$ for all n, then $1/y_n \to 1/M$.

We prove part (i) here. All the proofs are similar, and this part will illustrate the most important idea, which is that we need to choose things carefully to make everything work out. Here's what we mean: For every $\epsilon > 0$, we need to find N such that for $n \geq N$ we have $|(x_n + y_n) - (L + M)| < \epsilon$. We can make $|x_n - L| < \epsilon$ for n large enough, and we can make $|y_n - M| < \epsilon$ for n large enough, but if we add these together we get 2ϵ. You'll now see how we handle this problem.

Proof of (i).
Let $\epsilon > 0$. Since $\epsilon/2 > 0$ and $x_n \to L$, there exists N_1 such that $|x_n - L| < \epsilon/2$ for $n \geq N_1$. Again, since $\epsilon/2 > 0$ and $y_n \to M$, there exists N_2 such that $|y_n - M| < \epsilon/2$ for $n \geq N_2$. Let $N = \max\{N_1, N_2\}$. Then for $n \geq N$, we know that $n \geq N_1$ and $n \geq N_2$, so

$$
\begin{aligned}
|(x_n + y_n) &- (L + M)| \\
&= |(x_n - L) + (y_n - M)| \\
&\leq |x_n - L| + |y_n - M| \quad \text{(by the triangle inequality)} \\
&< \epsilon/2 + \epsilon/2 \quad\quad\quad\quad \text{(since } n \geq N_1 \text{ and } n \geq N_2 \text{).}
\end{aligned}
$$

Therefore, for all $\epsilon > 0$ there exists N such that for $n \geq N$ we have $|(x_n + y_n) - (L + M)| < \epsilon$, as desired. ∎

Solutions to Exercises

Solution to Exercise (19.1).
The answers are (in this order): $3/5$, 0, and -1.

Solution to Exercise (19.4).
Normally when we write up our solution we will include the proof and not our work on devising a plan. But one more careful example here will certainly be useful. So here's our plan: We guess that this converges to 0, so we need to show that for all $\epsilon > 0$, there exists N such that $|(2n+4)/(n^2+n+1)| < \epsilon$ for all $n \geq N$. Now both numerator and denominator are a bit complicated. For the denominator, we need to find something smaller and simpler than n^2+n+1 involving the highest order term n^2. So for this part, we note that $n^2+n+1 > n^2$. Now for the numerator, we need to find something larger and simpler than $2n + 4$ involving the highest-order term n. For $n \geq 1$, since $4 \leq 4n$, we see that $2n + 4 \leq 2n + 4n = 6n$. Putting this together, for $n \geq 1$, we have

$$
\left| \frac{2n + 4}{n^2 + n + 1} \right| \leq \frac{6n}{n^2} = \frac{6}{n}.
$$

So, if we make $6/n < \epsilon$, we should be able to complete the proof. Thus, we'll choose $N = 7/\epsilon$.

Proof.
If $\epsilon > 0$, we choose $N = 7/\epsilon$. Then for $n \geq N$, we have

$$\left| \frac{2n+4}{n^2+n+1} - 0 \right| \leq \frac{6n}{n^2} = \frac{6}{n} \leq \frac{6}{N} = \frac{6\epsilon}{7} < \epsilon,$$

where the first inequality follows since $n \geq 1$ and, consequently, $2n + 4 \leq 6n$. Thus $\big((2n+4)/(n^2+n+1)\big)$ converges to 0. ∎

Solution to Exercise (19.5).
A sequence (x_n) diverges if for every real number L there exists $\epsilon > 0$ such that for all $N \in \mathbb{R}$ there exists $n \geq N$ with $|x_n - L| \geq \epsilon$.

Solution to Exercise (19.6).

Proof.
Let L be a real number, and let $\epsilon = 1/2$. We break this into two cases. First suppose that $L < 0$. Let $N \in \mathbb{R}$ and choose n to be an even integer satisfying $n \geq N$. Then $|x_n - L| = |1 - L| > 1 > \epsilon$. Now if $L \geq 0$, then for $N \in \mathbb{R}$ choose an odd integer with $n \geq N$. It follows that $|x_n - L| = |-1 - L| = |1 + L| \geq 1 > \epsilon$. Therefore (x_n) diverges. ∎

Problems

Problem♭ 19.1.
We used the following several times in this chapter: Let $x, y, z \in \mathbb{R}$. Then $|x - y| \leq |x - z| + |y - z|$. Prove this statement.

Problem 19.2.
Let a and δ be real numbers with $\delta > 0$. Show that for all real numbers x, we have $|x - a| < \delta$ if and only if $a - \delta < x < a + \delta$.

Problem 19.3.
For each of the following, guess the limit and then prove (using the definition of convergence) that your guess is correct:

(a) $\displaystyle \lim_{n \to \infty} \frac{1}{3n}$;

(b) $\lim\limits_{n\to\infty} \dfrac{1}{\sqrt{n}}$;

(c) $\lim\limits_{n\to\infty} \dfrac{1}{\sqrt{n+7}}$;

(d) $\lim\limits_{n\to\infty} \dfrac{n^2+4}{n^2}$;

(e) $\lim\limits_{n\to\infty} \dfrac{2n+1}{n+2}$;

(f) $\lim\limits_{n\to\infty} \dfrac{3}{n!}$;

(g) $\lim\limits_{n\to\infty} \dfrac{1}{(n+7)!}$;

(h) $\lim\limits_{n\to\infty} \dfrac{3n^2+1}{4n^2+n+2}$.

Problem 19.4.
Prove Theorem 19.9, part (*ii*).

Problem 19.5.

(a) Suppose that (x_n) and (y_n) are sequences and $0 \le x_n \le y_n$ for all positive integers n. Show that if $y_n \to 0$, then $x_n \to 0$.

(b) Suppose that (x_n) and (y_n) are sequences and $-y_n \le x_n \le y_n$ for all positive integers n. Show that if $y_n \to 0$, then $x_n \to 0$.

(c) Find $\lim\limits_{n\to\infty} \dfrac{\sin^2 n}{n}$ and $\lim\limits_{n\to\infty} \dfrac{(-1)^n}{n^2+1}$, and prove that your answers are correct.

Problem 19.6.
Redo Problem 19.3, parts (a), (c), (d), (e), (g) and (h) using theorems in this chapter or Problem 19.5.

Problem 19.7.

(a) Show that for every sequence (x_n) we have $0 \le |x_n|+x_n \le 2|x_n|$.

(b) Prove that if $|x_n| \to 0$, then $x_n \to 0$. (See Problem 19.5.)

(c) If (x_n) is a sequence such that $|x_n| \to 1$, must $x_n \to 1$?

Problem 19.8.
Prove Theorem 19.9, part (*iii*). (Hint: You may want to use Theorem 19.8.)

Problem 19.9.

The proof of Theorem 19.9, part (iv) is outlined below.

(a) Prove that for real numbers y and M, if $|y - M| \leq |M|/2$, then $|y| \geq |M|/2$. (You might wish to use the lower triangle inequality to establish this implication.)

(b) Let (y_n) be a sequence of nonzero real numbers, and suppose that $y_n \to M$, where $M \neq 0$. Prove that if $0 < \epsilon < |M|/2$, then there exists N such that for $n \geq N$ if $|y_n - M| < \epsilon$, then $|(M - y_n)/(My_n)| \leq (2/M^2)|M - y_n|$.

(c) Prove Theorem 19.9, part (iv).

Problem 19.10.

Let a be a real number satisfying $0 < a < 1$.

(a) Show that there exists a real number x such that $x > 0$ and $a = 1/(1 + x)$.

(b) Show that $a^n \leq 1/(1 + nx)$ for all $n \in \mathbb{N}$. (You will need to do Problem 17.6 if you have not already done it.)

(c) Show carefully that $1/(1 + nx) \to 0$.

(d) Show that $a^n \to 0$. (You will want to do Problem 19.5 if you have not already done it.)

(e) Show that if $x_n = \underbrace{0.999...9}_{n \ 9's}$, then $x_n \to 1$.

Problem 19.11.

Let $x_n = F_{n+1}/F_n$, where F_n denotes the nth Fibonacci number. In Problem 18.12 we showed that $x_n = 1 + 1/x_{n-1}$. Assume further that there exists a nonzero real number L such that $x_n \to L$. Explain why $1/x_n \to 1/L$ and use these facts to compute L.

The number L is the golden ratio, which appears frequently in architecture and in nature. The Greeks, and others, felt (and still feel) that rectangles with sides in golden ratio are the most beautiful.

Problem 19.12.

(An exercise in reading and writing.)

(a) Read the proof below until you understand it. Mathematicians often read proofs many times, and you may have to do so with this one.

Theorem 19.10.
Every increasing bounded sequence converges to its supremum.

Proof.
Let $l = \sup(x_n)$, and let $\epsilon > 0$. Since $l - \epsilon$ is not an upper bound, there exists N such that $x_N > l - \epsilon$. We have assumed that (x_n) is an increasing sequence. Therefore, if $n \geq N$, we know that $x_n \geq x_N > l - \epsilon$. Since $x_n \leq l$ for all n, for $n \geq N$ we have $l - \epsilon < x_n < l + \epsilon$, and thus $|x_n - l| < \epsilon$. Therefore, the sequence (x_n) converges to l. ∎

(b) Use the ideas in the proof above to prove Theorem 19.11.

Theorem 19.11.
Every decreasing bounded sequence converges to its infimum.

(c) Can you find another proof of Theorem 19.11, this time using the statement of Theorem 19.10 rather than its proof?

Problem 19.13.
Use the theorems of Problem 19.12 for the following.
(a) Show that $(n!/n^n)$ converges.
(b) Let (x_n) be a bounded sequence. For each positive integer n, let $s_n = \sup\{x_m : m \geq n\}$. Prove that (s_n) converges. The limit of (s_n) is called the **limit superior** of (x_n), or simply $\limsup x_n$. What is the $\limsup(-1)^n$?
(c) Let (x_n) be a bounded sequence. Let $t_n = \inf\{x_m : m \geq n\}$. Prove that (t_n) converges. The limit of (t_n) is called the **limit inferior** of (x_n), or simply $\liminf x_n$.

Problem 19.14.
Sequences afford an excellent opportunity to practice everything you have learned. That's what you'll do in this problem: For each of the definitions below, do the following.
1. Read the definition.
2. Try to find an example of something that illustrates the definition.
3. Try to find an example of something that does not satisfy the defining conditions.
4. Write the definition in symbols.

5. Negate the definition.
   (a) A sequence (x_n) of real numbers is said to **diverge to infinity** (written $x_n \to \infty$) if for every $M \in \mathbb{R}$ there exists a real number N such that $n \geq N$ implies $x_n > M$.
   (b) A sequence is **monotone** if it is either increasing or decreasing.
   (c) A sequence (x_n) is a **Cauchy sequence** if for all $\epsilon > 0$ there exists a real number N such that $n, m \geq N$ implies that $|x_n - x_m| < \epsilon$.

   In addition, for this one, pretend you were talking to a high school student who loves mathematics and just *has* to know what a Cauchy sequence is but has never heard of ϵ and N. What would you tell him or her?

Problem 19.15.
Consider the following definition.

Let $(x_n)_{n=1}^\infty$ be a sequence of real numbers and let n_k be a sequence of positive integers with $n_1 < n_2 < n_3 < \cdots$. The sequence (x_{n_k}) is called a **subsequence** of (x_n). For this problem, do all the things you did in Problem 19.14, plus (a), (b), and (c) below.

(a) This definition says that when you choose a subsequence, you must do two things: you need to list all the x_n in order of appearance, and then you obtain the x_{n_k} by choosing one element after the other from your list, making sure that the term you choose comes after the one you just chose. How does the definition tell you that you must choose from this list? How does it tell you that you must choose in order of increasing appearance?
(b) Let $x_n = 1/n$. Is (x_n) a subsequence of itself? If $y_n = x_{2n}$, give a formula for y_n in terms of n. If $z_n = x_{n+4}$, give a formula for z_n in terms of n.
(c) How can you tell that the sequence (w_n) given by $w_n = (-1)^n/n$ is not a subsequence of (x_n) as defined in (b)?

Problem 19.16.
In Problem 19.14, part (c) we defined a Cauchy sequence. Show that every convergent sequence is a Cauchy sequence.

20

CHAPTER

Equivalent Sets

If you were asked how many people are in your class, you would do the natural thing and count them. Thinking about this carefully, we see that what you are doing is assigning each person in the room one and only one number. If we asked whether or not there are more people in this class than were in your high school geometry class, you could certainly answer that question by comparing the two numbers.

Now suppose we look at the positive integers and the natural numbers. Which set has more elements? What should that even mean? This is a more difficult question to answer correctly than you might think. The following mathematical folktale (often attributed to Hilbert) illustrates the problem and a solution.

Suppose there is a hotel with infinitely many rooms. The hotel is completely booked when the coach of a Davis Cup team arrives. The clever manager accommodates her by moving all of the other guests to the room numbered one higher than the room they previously occupied, which clears the first room for the coach. Then the four members of the team arrive. Each must have his own room, of course. The very clever manager moves everyone up four rooms, making enough room for the four athletes. Finally, the team's infinitely many fans arrive (this happens all the time at really good hotels). The very, very clever manager accommodates all guests by moving the residents of room number n to the room with number $2n$. Now all the new people can go in the odd-numbered rooms.

An interesting commentary of this problem was given by Smilla in the book *Smilla's Sense of Snow*. She says, "What delights me about this story is that everyone involved, the guests and the owner, accept it as perfectly natural to carry out an infinite number of operations so that one guest can have peace and quiet in a room of his own. That is a great tribute to solitude."[1] [42, p. 11].

What does "Hotel Infinity" really show us? It is our aim in the next few sections to discuss and answer these questions.

To make precise what it means for two sets (even two infinite sets) to have the same number of elements, we need a definition. We say that a set A is **equivalent** to a set B if there exists a bijection $f : A \to B$. We write $A \approx B$ for A is equivalent to B. (Other authors use the words equipotent or equinumerous.)

You actually know a lot about this concept from previous chapters (particularly Chapter 14). The next result summarizes information we already have.

Theorem 20.1.
Let X be a nonempty set. Equivalence between subsets of X, as defined by $A \approx B$ above, is an equivalence relation on $\mathcal{P}(X)$.

A proof is outlined in the directions to Problem 20.5.

Example 20.2.
Show that the open interval $(0, 1)$ is equivalent to the open interval $(0, 3)$.

Proof.
Define a function $f : (0, 1) \to (0, 3)$ by $f(x) = 3x$. We leave it to you to show that this function is bijective. Thus $(0, 1) \approx (0, 3)$. ∎

---

[1] From *Smilla's Sense of Snow* by Peter Høeg, published by Farrar, Straus and Giroux, 1993. Reprinted by permission of Farrar, Straus and Giroux. Published in the UK by Harvill Press. Reprinted by permission of The Random House Group Ltd.

Since the relation $\approx$ is symmetric, we will usually say that A and B are equivalent, rather than A is equivalent to B. This concept allows us to give a precise definition of a finite set. We say that a set S is **finite** if either $S = \emptyset$ or if S is equivalent to the set $\{1, 2, 3, \ldots, n\}$ for some positive integer n. Thus, to prove that a nonempty set is finite, we need to find a bijection between S and a set $\{1, 2, 3, \ldots, n\}$ for some $n \in \mathbb{Z}^+$. Since the relation is symmetric, either set can serve as the domain. The bijection is the mathematical analog of what we usually describe as counting. A set is said to be **infinite** if it is not finite.

Example 20.3.
By negating the definition of finite, say what it means for a set to be infinite.

A set S is infinite if it is nonempty and for every positive integer n there does not exist a bijection from S to the subset $\{1, 2, 3, \ldots, n\}$.
○

What are some examples of finite sets? By our definition, every set of the form $\{1, 2, 3 \ldots, n\}$ is a finite set. What about a set like $\{2, 4, 6\}$? It certainly feels finite, but to prove that it is finite we would have to construct a bijective function. It is easy here; the function $f : \{1, 2, 3\} \rightarrow \{2, 4, 6\}$ defined by $f(n) = 2n$ certainly works. You could also define a function g on $\{1, 2, 3\}$ by $g(1) = 4, g(2) = 6$, and $g(3) = 2$. In fact, if our set has more than one element, then there is more than one choice for the bijection.

Exercise 20.4.
Show that the set $\{6, 8, 10, 14\}$ is finite.
○

The rest of this chapter is a paraphrasing of much of the work we did in Chapters 13 through 16. We isolate the important ideas below and we guide you through the proofs, but you have all the techniques to prove everything yourself.

Theorem 20.5.
The sets $\mathbb{Z}$ and $\mathbb{N}$ are equivalent.

This theorem is really the essence of the story behind "Hotel Infinity." An infinite set, $\mathbb{Z}$, can have a proper subset, $\mathbb{N}$, that has the "same number of elements" in it!

Proof.
Define $f : \mathbb{Z} \rightarrow \mathbb{N}$ explicitly as follows:

$$f(x) = \begin{cases} 2x & \text{if } x \geq 0 \\ -(1 + 2x) & \text{otherwise} \end{cases}.$$

In Example 14.7, we showed that this function is bijective. From this we conclude that $\mathbb{Z} \approx \mathbb{N}$. ∎

The same techniques that were used to prove Theorem 20.5 can be used to prove the next theorem.

Theorem 20.6.
Let A, B, C, and D be nonempty sets. Suppose that $A \cap B = \emptyset$, $C \cap D = \emptyset$, $A \approx C$, and $B \approx D$. Then $A \cup B \approx C \cup D$.

Outline of proof.
Since $A \approx C$, there exists a bijective function $f : A \rightarrow C$. Similarly, since $B \approx D$, there exists a bijective function $g : B \rightarrow D$. We define a function $H : A \cup B \rightarrow C \cup D$ in cases by

$$H(x) = \begin{cases} f(x) & \text{if } x \in A \\ g(x) & \text{if } x \in B \end{cases}.$$

We leave it to you to show that H is well-defined and bijective. ∎

Exercise 20.7.
What happens to our proof if $A \cap B \neq \emptyset$? If we do not assume that $A \cap B = \emptyset$, is the theorem still true? ○

For finite sets, Theorem 20.6 has an interesting consequence.

Corollary 20.8.
Let A and B be disjoint sets. If A and B are finite, then $A \cup B$ is finite.

The proof of Corollary 20.8 is left to you in Problem 20.9.

Since the definition of equivalent sets uses bijective functions, many of the theorems you have had will come in quite handy. We recall here the definition of restriction that we presented in Problem 15.20.

The definition of a function includes the specification of a domain and codomain. So if the domains of two functions are not the same, then the functions cannot possibly be equal. Sometimes, though, we like our function's definition, but we need to restrict the elements in the domain to members with certain properties. As soon as we change the domain, we must admit that we changed the function. This new function is closely related to the old one, and the notation should reflect that. That's the essence of the following definition: If $f : A \rightarrow B$ is a function, and $A_1 \subset A$, we define another function $F : A_1 \rightarrow B$ by $F(a) = f(a)$ for all $a \in A_1$. This function F is called the **restriction** of f to A_1 and is usually denoted $f|_{A_1}$.

Exercise 20.9.
With the notation of the last paragraph show that if f is one-to-one, then $f|_{A_1}$ is one-to-one. Conclude that if $f|_{A_1}$ is not one-to-one, then f is not one-to-one. ○

We'll use the restriction function in many of the proofs about finite and infinite sets.

Theorem 20.10.
Let n be a positive integer. Then every subset of $\{1, 2, 3, \ldots, n\}$ is finite.

Proof.
The proof will be by induction on n.

If $n = 1$, then our set is $\{1\}$, and there are only two subsets; $\{1\}$ and $\emptyset$. Therefore, the result holds if $n = 1$.

Our induction hypothesis states that every subset of $\{1, 2, \ldots, n\}$ is finite. We must show that every subset of $\{1, 2, \ldots, n, n + 1\}$ is finite. So consider the set $\{1, 2, \ldots, n, n + 1\}$, and a subset S of this set. If $S \subseteq \{1, 2, \ldots, n\}$, then S is finite by our induction hypothesis. Otherwise, $n + 1 \in S$. In this case, notice that $\{n + 1\}$ is a finite set. Since $S \setminus \{n + 1\} \subseteq \{1, \ldots, n\}$ we also know from the induction hypothesis that $S \setminus \{n+1\}$ is finite. Applying Corollary 20.8 we may conclude that $S = (S \setminus \{n+1\}) \cup \{n+1\}$ is finite, completing the proof. ∎

Corollary 20.11.
Let S be a finite set. Then every subset of S is finite.

Proof.
If S is empty, the result is clear. So suppose that S is nonempty, and let T denote a subset of S. Again, we may assume that T is nonempty.

Now, by our assumption there exists a positive integer n and a bijection $f : S \to \{1, 2, \ldots, n\}$. By Exercise 20.9 the restriction function, $f|_T$, is a bijective mapping from T onto a nonempty subset B of $\{1, 2, \ldots, n\}$. From Theorem 20.10, the set B is finite, and therefore there exists a positive integer m and a bijection $g : B \to \{1, 2, \ldots, m\}$. By Theorem 15.6, the composition $h = g \circ (f|_T)$ of the two bijective functions g and $f|_T$ is a bijection of T onto $\{1, 2, \ldots, m\}$. Thus T is finite, completing the proof. ∎

We assumed in Corollary 20.8 that the finite sets A and B were disjoint, but our intuition tells us that the union of two finite sets should be finite. How do we prove this? The idea is that the union of two sets can be expressed as the union of two disjoint sets: the intersection appears in the union twice, so to speak, so if we remove it once (from one of the sets) we haven't changed the union. That is the key to the next result.

Theorem 20.12.
The union of two finite sets is finite.

Proof.
Let A and B denote the two finite sets. Now you have already shown (in Problem 7.9) that

$$A \cup B = (A \setminus B) \cup B,$$

and it should be clear that these two sets are disjoint. By Corollary 20.11, the set $A \setminus B$ is finite. The set B is finite by assumption. Since these are two disjoint sets, we have written $A \cup B$ as the disjoint union of two finite sets and Corollary 20.8 now implies that $A \cup B$ is finite. ∎

Note that in the theorem above, we showed that a set is finite without exhibiting a specific bijection. We'll be building up many results that are useful, and we will not always go back to the definition to see how to prove things. This is a great plus—proofs become shorter, and sometimes prettier and more interesting. Of course, to use the theorems, you also have to know what they are!

We conclude this chapter with a useful exercise.

Exercise 20.13.
Use induction to prove the following. Let $m \in \mathbb{Z}^+$. If $A_1, A_2, \ldots, A_m$ are finite sets, then the union $\bigcup_{j=1}^{m} A_j$ is finite.

Solutions to Exercises

Solution to Exercise (20.7).
If A and B are not disjoint, the function H may not be well-defined. In addition, if A and B are not disjoint, the conclusion of the theorem may not hold. To see this, take finite sets with $A = B = \{1\}$, $C = \{2\}$ and $D = \{3\}$.

Solution to Exercise (20.9).
If x_1 and x_2 are elements of A_1, and $f|_{A_1}(x_1) = f|_{A_1}(x_2)$, then by the definition of $f|_{A_1}$ we know that $f(x_1) = f(x_2)$. Since we assume that f is one-to-one, we may conclude that $x_1 = x_2$, as desired. The final assertion is simply the contrapositive of what we just proved.

Solution to Exercise (20.13).
We will prove this statement by induction on m. For the base step $(m = 1)$, we get $\bigcup_{j=1}^{1} A_j = A_1$, which is finite by assumption.

Now assume that $\bigcup_{j=1}^{n} A_j$ is finite. We need to show that $\bigcup_{j=1}^{n+1} A_j$ is finite. But $\bigcup_{j=1}^{n+1} A_j = \left(\bigcup_{j=1}^{n} A_j \right) \cup A_{n+1}$. By the induction hypothesis, $\bigcup_{j=1}^{n} A_j$ is finite, and A_{n+1} is also assumed to be finite. By Theorem 20.12, the union of two finite sets is finite. Thus $\bigcup_{j=1}^{n+1} A_j$ is finite. The result now follows from the principle of mathematical induction.

Problems

Problem 20.1.
Show that the following intervals of real numbers are equivalent:
(a) $[0, 1]$ and $[0, 2]$;
(b) $[0, 1]$ and $[2, 5]$.

Problem 20.2.
Prove that $\{1, \ldots, 10\} \times \{1, \ldots, 15\}$ is finite using only the definition of a finite set; that is, write down the relevant bijection explicitly.

Problem 20.3.
Explain, in words, the difference between a "finite union of sets" and a "union of finite sets." Give examples of each that show these really are different.

Problem♮ 20.4.
(a) Show that the positive rationals $\mathbb{Q}^+$ and the negative rationals $\mathbb{Q}^-$ are equivalent.
(b) Show that the even and odd integers are equivalent.

Problem 20.5.
Prove Theorem 20.1 by doing all parts of the problem below.
(a) Write out in words what you must show to conclude that the relation $\approx$ is reflexive. Then prove it.
(b) Recall that the definition of symmetric has both a hypothesis and conclusion. Write out what the hypothesis $A \approx B$ means, and what you must show. Then prove it. Remember: if you have theorems that help you, you can use them.
(c) Recall that the definition of transitive has a hypothesis as well. Write out what it means when we say $A \approx B$ and $B \approx C$, and

what you need to prove. Then prove it. Remember: if you have theorems that help you, you can use them.

Problem 20.6.
  (a) Prove that $(0, 1) \approx \mathbb{R}$. (If you choose to use Problem 14.11, make sure you solve that problem too!)
  (b) Prove that $\mathbb{R} \approx \mathbb{R}^+$.

Problem 20.7.
  (a) Show that $\mathbb{Z} \approx 2\mathbb{Z}$.
  (b) Using theorems from this chapter (don't define functions!) show that $2\mathbb{Z} \approx \mathbb{N}$.

Problem 20.8.
Prove Theorem 20.6 working with the outline given in the text.

Problem 20.9.
  (a) Suppose that A and B are nonempty finite sets and $A \cap B = \emptyset$. Show that there exist integers n and m such that $A \approx \{1, 2, \ldots, n\}$ and $B \approx \{n + 1, \ldots, n + m\}$.
  (b) Prove Corollary 20.8.

Problem 20.10.
Prove Theorem 20.14 below. We suggest that you start by working Problem 15.14 if you have not already done so.

Theorem 20.14.
Let A, B, C, and D be nonempty sets with $A \approx C$ and $B \approx D$. Then $A \times B \approx C \times D$.

Problem 20.11.
Prove the following corollary of Theorem 20.14 above.

Corollary 20.15.
Let A and B be finite sets. Then $A \times B$ is a finite set.

Problem 20.12.
Let $\mathcal{A} = \{[a, b) : a, b \in \mathbb{R} \text{ and } a < b\}$ be the collection of bounded half-open intervals of real numbers.

(a) Prove that $\mathcal{A} \approx \mathbb{R} \times \mathbb{R}^+$.
(b) Prove that $\mathcal{A} \approx \mathbb{R} \times \mathbb{R}$.

21

CHAPTER

Finite Sets and an Infinite Set

We have proved that sets are finite, but we have not yet rigorously shown that a set is infinite. It is not as easy as you might think to show rigorously that something is infinite. We also do not have an exact notion of what it means for a finite set "to have n elements." Our proof of the former and the definition of the latter will depend on a principle known as the *pigeonhole principle*. Here is a problem that will serve as an introduction to this principle.

Theorem 21.1.
Suppose that n people (n $\geq$ 2) are at a party. Then there exist at least two people at the party who know the same number of people present.

First you need to know the rules. We will assume that no one knows him or herself. We will also assume that if x claims to know y, then y also knows x.

The idea behind the proof is this, and you can try it out at your next party. You will put n boxes on the board numbered 0 through $n-1$. Each person counts up the number of people he or she knows at the party. You ask them that number and write their name in the box with the same number. Note that each person's answer corresponds to exactly one of the boxes 0 through $n - 1$. The theorem claims that at least two people's names will end up in the same box.

Proof.

We imagine n boxes, numbered 0 through $n - 1$. For an integer m with $0 \leq m \leq n - 1$, box m contains the names of the people who know m people at the party.

We break this proof into two cases. First, suppose that there is someone at the party who doesn't know anyone. We'll call this party crasher Ms. X. Now if we pick some other party attendee, he doesn't know himself and, since Ms. X doesn't know him, he doesn't know Ms. X either. This implies that he knows at most $n - 2$ people at the party. The point is this: No one's name can be in the box labeled $n - 1$. This means that the names of the n people are in the $n - 1$ boxes labeled 0 through $n - 2$. Obviously then, there is a box with at least two names in it, indicating that two of those people know the same number of people and we are done in this case.

Now suppose that everyone knows at least one person at the party. Then no one's name can be in the box marked 0; everyone's name will be in one of the $n - 1$ boxes marked $1, \ldots, n - 1$. Once again we have n names in $n - 1$ boxes, and thus at least two must be in the same box. ∎

This theorem and its proof illustrate the idea behind the pigeonhole principle. In its popular form, the principle says that *if there are more pigeons than holes, then at least one hole is the home of more than one pigeon.* There are many wonderful applications of the pigeonhole principle (and many can be found at the web site [10] under algebra).

We now turn to the more precise statement of the pigeonhole principle and its proof. The principle is attributed to Peter Gustav Lejeune Dirichlet and is also known as the Dirichlet principle or the Dirichlet drawer principle. (There are other Dirichlet principles; see the Spotlight: Minimum or Infimum in Chapter 16.) Curiously, this intuitively obvious principle has a rather intricate proof.

Theorem 21.2 (Pigeonhole principle).

Let m and n be positive integers with m > n, and let f be a map satisfying
$f : \{1, \ldots, m\} \to \{1, \ldots, n\}$. *Then f is not one-to-one.*

Proof.

We will prove this theorem by induction on n. The assertion that we will prove is: For every integer m such that $m > n$, if $f : \{1, \ldots, m\} \to \{1, \ldots, n\}$, then f is not one-to-one.

For the base case, $n = 1$. Our assumption that $m > n$ implies that $m > 1$. For every map $f : \{1, \ldots, m\} \to \{1\}$, it is clear that $f(1) = 1 = f(m)$. Since $1 \neq m$, we may conclude that f is not one-to-one, completing the base step.

For the induction step, let us assume that for an $n \geq 1$, if m is an integer greater than n and f is a function $f : \{1, \ldots, m\} \to \{1, \ldots, n\}$, then f is not one-to-one. We will show that if $m > n + 1$ and $f : \{1, \ldots, m\} \to \{1, \ldots, n+1\}$, then f is not one-to-one.

So let us suppose that $m > n + 1$ and we have a map

$$f : \{1, \ldots, m\} \to \{1, \ldots, n+1\}.$$

There are three cases to consider: Either f maps nothing to $n + 1$, more than one element to $n + 1$, or exactly one element to $n + 1$.

For the first case, if $n + 1 \notin \mathrm{ran}(f)$, then f actually defines a map $f : \{1, \ldots, m\} \to \{1, \ldots, n\}$. Since $m > n + 1 > n$, our induction hypothesis tells us that f is not one-to-one, and we are done in this case.

For the second case, suppose there exist $j, k \in \{1, \ldots, m\}$ with $j \neq k$, and $f(j) = n + 1 = f(k)$. Then f is not one-to-one, and we are done in this case, too.

For the last case, assume that $j \in \{1, \ldots, m\}$ is the only integer for which $f(j) = n + 1$. We now define the function $g : \{1, \ldots, m\} \to \{1, \ldots, m\}$ that interchanges m with j and leaves all other elements of $\{1, \ldots, m\}$ fixed:

$$g(k) = \begin{cases} k & \text{if } k \neq j, m \\ j & \text{if } k = m \\ m & \text{if } k = j \end{cases}.$$

Then g is clearly one-to-one. Furthermore, since j is the only integer that f maps to $n + 1$ we know that $(f \circ g)(k) = f(g(k)) = n + 1$ if and only if $g(k) = j$. Since g is one-to-one, this happens if and only if $k = m$. Thus $(f \circ g)|_{\{1,...,m-1\}}$ maps $\{1, \ldots, m-1\}$ to $\{1, \ldots, n\}$. Since we assume that $m > n+1$, we know that $m-1 > n$. Our induction hypothesis now applies, and we conclude that $(f \circ g)|_{\{1,...,m-1\}}$ is not one-to-one. By Exercise 20.9 $f \circ g$ is not one-to-one. Since g is one-to-one, by Theorem 15.7 f is not one-to-one in this case either.

By the induction principle, if $m, n \in \mathbb{Z}^+$ and $m > n$, then no function $f : \{1, \ldots, m\} \to \{1, \ldots, n\}$ is one-to-one. ∎

The proof of the pigeonhole principle summarizes much of what you learned: mathematical induction, proof in cases, and one-to-one functions.

We are now in a position to prove that a set is infinite.

Theorem 21.3.
The set $\mathbb{N}$ *is infinite.*

Proof.
Suppose to the contrary that $\mathbb{N}$ is finite. Since $\mathbb{N} \neq \emptyset$ there exists an integer m and a one-to-one mapping, g, of $\mathbb{N}$ onto $\{1, 2, \ldots, m\}$. Now $\{1, 2, \ldots, m+1\} \subseteq \mathbb{N}$, so we may consider the restriction $g|_{\{1,2,...,m+1\}} : \{1, 2, \ldots, m+1\} \to \{1, 2, \ldots, m\}$. The pigeonhole principle (Theorem 21.2) implies that $g|_{\{1,2,...,m+1\}}$ is not one-to-one. This, in turn, implies (as you surely showed in Exercise 20.9) that g is not one-to-one, contradicting our choice of g. Therefore, it must be the case that $\mathbb{N}$ is infinite. ∎

Exercise 21.4.
Prove that $\mathbb{Z}$ is infinite. ○

The next exercise is similar to the one above, but requires more work.

Exercise 21.5.
Prove that if X is an infinite set, then the power set of X is infinite. ○

We carefully defined what it means for a set to be finite, but so far we have not described what it means for a set to "have n elements." After the following theorem, we will be ready to do that.

Theorem 21.6.

Let A be a nonempty finite set. There is a unique positive integer n such that $A \approx \{1, \ldots, n\}$.

Before we begin, note that there are two things to show: there exists a positive integer n with certain properties and that there is only one such integer. One of these should be easy. Which one?

Proof.

The existence of some $n \in \mathbb{Z}^+$ such that there is a bijection $f : A \to \{1, \ldots, n\}$ is guaranteed by the definition of a nonempty finite set. So all we have to do is show that there is no other $m \in \mathbb{Z}^+$ with an associated bijection $g : A \to \{1, \ldots, m\}$.

Suppose to the contrary that there does exist such a positive integer m with $m \neq n$, and bijective function g. Since $m \neq n$, one of these integers must be larger than the other, so we assume that $m > n$. Since g is a bijection, it has an inverse. Composing the two bijective functions f and g^{-1}, we obtain a bijective function $f \circ g^{-1} : \{1, \ldots, m\} \to \{1, \ldots, n\}$. But this contradicts the pigeonhole principle, and we conclude that n is unique. ∎

The integer n in the above theorem is exactly what we mean by the "size of A" or "number of elements in A." We will say that the **cardinality** of a finite set A is 0 if A is empty and n if $A \approx \{1, \ldots, n\}$. In symbols, we write $|A| = n$, where $n \in \mathbb{N}$. (Why couldn't we define the cardinality of a set before we proved the theorem?) While it is possible to define cardinality in more generality, we have only defined it in the case that A is finite.

Solutions to Exercises

Solution to Exercise (21.4).

Proof.
We know that $\mathbb{N} \subset \mathbb{Z}$, and Theorem 21.3 tells us that $\mathbb{N}$ is infinite. Since Corollary 20.11 says that every subset of a finite set is finite, our set $\mathbb{Z}$ must be infinite. ∎

Solution to Exercise (21.5).

Proof.
Define a map $f : X \rightarrow \mathcal{P}(X)$ by $f(x) = \{x\}$ for all $x \in X$. It is clear that f is well-defined and one-to-one. Therefore, f maps X onto a subset, $\mathcal{A}$, of $\mathcal{P}(X)$. Since X is infinite and f is a bijection of X onto $\mathcal{A}$, we know that $\mathcal{A}$ must be infinite as well. Thus, $\mathcal{P}(X)$ contains an infinite subset, and we may use Corollary 20.11 to conclude that $\mathcal{P}(X)$ is infinite. ∎

Problems

Problem 21.1.
Consider the story of n people at a party in Theorem 21.1. Suppose someone else has a rival party the same evening, and no one can attend both. Someone takes a picture of the people at the rival party and shows it to everyone at your party. Your party isn't that much fun, so you each look at the picture and say how many people you know at the other party. No one says the same number. What can you conclude about the number of people attending the other party?

Problem 21.2.
  (a) Suppose there are 15 people in a class. Show that two people must be born in the same month.
  (b) A conductor has just taken on a new job in a small town where he has five trumpet players in his orchestra. He has a concert every other evening for his first year. Traditionally the players

are seated from left to right in order of decreasing musical ability. The conductor does not want to offend the players, so he has decided to seat them differently at each performance. Can he do it? Why or why not?

Problem 21.3.
Suppose 21 numbers are chosen from the set $\{1, \ldots, 40\}$. Show that among the chosen numbers there are (at least) two of them, n and m, such that $n - m = 1$.

Problem 21.4.
Let S be a region in the plane bounded by a square with sides of length two. Prove that if we put five points in S, there exist (at least) two of these points that are at most a distance of $\sqrt{2}$ apart.

Problem 21.5.
Let $n \in \mathbb{N}$ and let $f : \{1, 2, \ldots, 2n+1\} \to \{1, 2, \ldots, 2n+1\}$ be a bijective function. Prove that for some odd integer $k \in \{1, 2, \ldots, 2n + 1\}$, the integer $f(k)$ is also odd.

Problem 21.6.
Prove the following alternate form of the pigeonhole principle.

Let A and B be nonempty finite sets, and suppose that $|A| > |B|$. If $f : A \to B$ is a function, then f is not one-to-one.

Problem♭ 21.7.
Show that $\mathbb{Q}$ is infinite.

Problem 21.8.
Using only the definition of finite and the pigeonhole principle, prove that $\mathbb{R}$ is infinite.

Problem 21.9.
Let A be a set, and suppose that B is an infinite subset of A. Show that A must be infinite.

Problem 21.10.
Suppose that A is an infinite set, B is a finite set and $f : A \to B$ is a function. Show that there exists $b \in B$ such that $f^{-1}(\{b\})$ is infinite.

Problem 21.11.

Let X be an infinite set, and A and B be finite subsets of X. Answer each of the following, giving reasons for your answers:

(a) Is $A \cap B$ finite or infinite?

(b) Is $A \setminus B$ finite or infinite?

(c) Is $X \setminus A$ finite or infinite?

(d) Is $A \cup B$ finite or infinite?

(e) If $f : A \to X$ is a one-to-one function, is $f(A)$ finite or infinite?

Problem 21.12.

Let A, B, and C be finite sets.

(a) Recall that we showed that if A and B are disjoint, then $A \cup B$ is finite. Look over the proof outlined in Problem 20.9 and determine $|A \cup B|$ in terms of $|A|$ and $|B|$, assuming that A and B are disjoint.

(b) Now suppose that A and B are not disjoint. Show that $|A \cup B| = |A| + |B| - |A \cap B|$.

(c) Find a formula that works for three sets A, B, and C. (You don't need to prove that your formula works.)

Problem 21.13.

(a) Suppose that A and B are finite sets with $|A| = m$ and $|B| = n$. In Problem 20.11 you showed that if A and B are finite, then $A \times B$ is finite. Look over the proof and determine $|A \times B|$ in terms of m and n.

(b) Suppose that $A_1, A_2, \ldots, A_k$ are finite sets. Guess a formula for the cardinality of $A_1 \times A_2 \times \cdots \times A_k$ (in terms of $|A_1|, |A_2|, \ldots,$ and $|A_k|$). Prove that your formula is correct.

Problem 21.14.

Each of the problems below is an application of one of the counting principles given in Problems 21.12 and 21.13. Decide which part of that problem applies, and use it to answer the problem.

(a) Thirty second graders, twenty-five third graders, and fifteen fourth graders entered an art contest. Three prizes were awarded, one for each grade. In how many ways can the prizes be awarded to three of the children? (Don't forget to say which formula from Problem 21.12 or 21.13 applies.)

(b) Suppose that there are 100 people in a room. Of these 55 are men, 33 are Swiss, 10 are Swiss males. How many are Swiss or male (or both)? (Don't forget to say which formula from Problem 21.12 or 21.13 applies.)

The rest of the problems are interrelated. If you can't see how to do the problem you are working on, look at the results from the previous problems and Problem 21.12.

Problem 21.15.
Let A be a nonempty finite set with $|A| = n$ and let $a \in A$. Prove that $A \setminus \{a\}$ is finite and $|A \setminus \{a\}| = n - 1$.

Problem 21.16.
(a) Suppose that A is a finite set and $B \subseteq A$. We showed that B is finite. Show that $|B| \leq |A|$.
(b) Suppose that A is a finite set and $B \subseteq A$. Show that if $B \neq A$, then $|B| < |A|$.
(c) Show that if two finite sets A and B satisfy $B \subseteq A$ and $|A| \leq |B|$, then $A = B$.

Problem 21.17.
Suppose that A and B are finite sets and $f : A \to B$ is one-to-one. Show that $|A| \leq |B|$.

Problem 21.18.
Let A and B be sets with A finite. Let $f : A \to B$. Prove that $|\operatorname{ran}(f)| \leq |A|$.

Problem 21.19.
Let A be a finite set. Show that a function $f : A \to A$ is one-to-one if and only if it is onto. Is this still true if A is infinite?

22

CHAPTER

Countable and Uncountable Sets

I see it but I do not believe it. —Georg Cantor[48, p. 997]

Having mastered finite sets, we now turn to understanding the infinite. We know that $\mathbb{N}$ is infinite, and we know that $\mathbb{Q}$ is infinite (see Problem 21.7). Are they equivalent? In some sense, we can count $\mathbb{N}$ and it may feel as though we cannot count $\mathbb{Q}$—that is, as though we cannot list a first element, second element, third element, and so on. However, we shall see that $\mathbb{Q}$ and $\mathbb{N}$ are, in fact, equivalent.

An infinite set A is said to be **countably infinite** if $A \approx \mathbb{N}$. In Chapter 20 we showed that $\mathbb{Z} \approx \mathbb{N}$ and $2\mathbb{Z} \approx \mathbb{N}$, so these also are countably infinite. It is also easy to show that $\mathbb{Z}^+ = \mathbb{N} \setminus \{0\} \approx \mathbb{N}$. A set is **countable** if it is either finite or countably infinite. A set is said to be **uncountable** if it is not countable. Note that if we only know that a set is countable we don't necessarily know if it is finite or infinite. If we have an infinite countable set, it automatically is equivalent to $\mathbb{N}$.

Exercise 22.1.
Let A and B be two countably infinite sets. Prove that there is a bijection of A onto B.

271

Theorem 22.2.
Every subset of $\mathbb{N}$ *is countable.*

The proof of this theorem will be presented as an exercise in reading mathematics.

Exercise 22.3.
This is an exercise in reading a proof. We ask that as you read you pretend that the set T (appearing in the proof) is the set of prime numbers. Of course, you are not allowed to pick a particular subset, call it T, and conclude that you have proved the theorem. However, for the purpose of understanding someone else's proof, this might be quite helpful. We will call this set the demo, and denote it by T. Whenever you see (?) you should figure out what happens for this set. If at the end you remain largely unsatisfied, pick another set for T and try again. No matter what, you'll understand more of the proof than if you hadn't tried anything at all. So read the proof, think about the question marks and then answer the set of questions provided below.

1. What's t_1 in the demo?
2. Find $f(1), \ldots, f(5)$ in the demo. Remove these from T. What remains in our demo? What is $f(6)$ in the demo set T?
3. For one-to-one, we thought we had to show that if $f(i) = f(j)$, then $i = j$. What's going on in this proof?
4. Say $s = 17$. What gets mapped to s? ○

The idea of the proof is the following: If T is an infinite subset of $\mathbb{N}$, we will construct a function $f : \mathbb{Z}^+ \to T$ recursively, listing the elements of T in increasing order. Since the list is strictly increasing, f will be one-to-one. And since the list is constructed to look at what has been chosen and then move on to the next largest number in T, it will also be onto. We now make these ideas precise.

Note that the (?)'s in this proof refer to Exercise 22.3. Once you have worked through the exercise, you should be able to read through the proof, ignoring the symbol (?) as you read.

Proof.
Let T denote a subset of $\mathbb{N}$. If T is finite, it is countable and we are done.(?) We suppose, then, that T is infinite and show that it is countably infinite. We now explicitly construct a bijection $f : \mathbb{Z}^+ \to T$. Since $\mathbb{N} \approx \mathbb{Z}^+$, as pointed out above, this will establish that T is countable.(?)

Since T is a nonempty subset of $\mathbb{N}$, the well-ordering principle implies that T has a least element, which we will call t_1.(?) Define $f(1) = t_1$.(?) Now suppose that we have defined $f(1), \ldots, f(n)$ and define $f(n+1)$ as follows. Since T is not finite, $T \setminus \{f(1), \ldots, f(n)\}$ is nonempty.(?) By the well-ordering principle, we may choose the least element of this set, which we denote by t_{n+1}.(?) Now $f(n+1)$ may be defined by $f(n+1) = t_{n+1}$.

We first note that f is defined for every positive integer and there is a unique value associated with each positive integer; in other words, f is well-defined.

We will now show that the function f defined above is one-to-one. So let i and j be positive integers with $i \neq j$. We will show that $f(i) \neq f(j)$. Since $i \neq j$, one of them must be strictly smaller than the other, say $i < j$. Now f was constructed to make $t_1 < t_2 < \cdots < t_n < t_{n+1}$. It follows that $t_i < t_j$, and consequently $f(i) \neq f(j)$. Thus f is one-to-one.

Next we'll show that f is onto. If $s \in T$, consider the set $T_1 = \{t \in T : t < s\}$.(?) Then T_1 is finite (?), so there exists $n \in \mathbb{Z}^+$ such that $|T_1| = n$. Thus, there are n elements in T_1 and these have been enumerated by f as $f(1) = t_1, \ldots, f(n) = t_n$. Now $s \in T \setminus T_1$ and anything that is in T and less than s is in T_1, so s is also the least element of the set $T \setminus T_1 = T \setminus \{f(1), \ldots, f(n)\}$.(?) Hence $f(n+1) = s$. Therefore, f is onto. Since f is a bijection, T is countable. ∎

As you read the theorems and corollaries below, think about whether or not you know a corresponding result for finite sets. If so, what was the proof? Do the ideas from those proofs help you here? Why or why not? We leave the corollaries for you to prove in Problems 22.6 and 22.7.

Corollary 22.4.
Every subset of a countable set is countable.

Remember that when we say that a set is countable, we mean that it is finite or countably infinite. The next exercise will often allow you to handle both cases at once.

Exercise 22.5.
Prove that a nonempty set A is countable if and only if there exists a one-to-one function $f : A \to \mathbb{N}$.

Theorem 22.6.
Suppose that A and B are countable. Then $A \cup B$ is countable.

Our proof begins with something we have used several times before.

Proof.
If $A \subseteq B$ or $B \subseteq A$, the result is clear. So suppose that $A \setminus B$ and $B \setminus A$ are both nonempty. Now note that $A \cup B = A \cup (B \setminus A)$ and $A \cap (B \setminus A) = \emptyset$. Since $B \setminus A \subseteq B$, Corollary 22.4 implies that $B \setminus A$ is countable. Further, A and $B \setminus A$ are countable, so by Exercise 22.5 there exist one-to-one functions f and g such that $f : A \to \mathbb{N}$ and $g : B \setminus A \to \mathbb{N}$. Define $H : A \cup B \to \mathbb{N}$ by

$$H(x) = \begin{cases} 2f(x) & \text{if } x \in A \\ 2g(x) + 1 & \text{if } x \in B \setminus A \end{cases} .$$

You can check (as you have many times before) that H is well-defined and one-to-one. Using Exercise 22.5 once again, we conclude that $A \cup B$ is countable. ∎

Corollary 22.7.
The union of finitely many countable sets is countable.

In the next theorem, we want to show that $\mathbb{N} \times \mathbb{N}$ is equivalent to $\mathbb{N}$. It is oh, so tempting to go to the definition and try to define a function that is a bijection from our set onto $\mathbb{N}$; after all, that is the definition of equivalence. But if we do everything using definitions, we will not be taking advantage of the useful body of mathematics we have proved thus far, and we will have to re-prove everything we have done. Some of it was quite difficult to prove! Life will be much easier if we think about the theorems we have proved already and see when and how we can use them.

Theorem 22.8.
The set $\mathbb{N} \times \mathbb{N}$ is countable.

Proof.
We show that $\mathbb{N} \times \mathbb{N}$ is countable by defining a function $f : \mathbb{N} \times \mathbb{N} \to \mathbb{N}$ explicitly. So, define $f(n, m) = 2^n 3^m$ for all $(n, m) \in \mathbb{N} \times \mathbb{N}$. (Note that this function is not onto, since a number like 7 is not in the range. Therefore, we will not try to show that f is a bijection between $\mathbb{N} \times \mathbb{N}$ and $\mathbb{N}$.) Since the prime factorization of a natural number is unique, the function is one-to-one. Thus we have a one-to-one mapping $f : \mathbb{N} \times \mathbb{N} \to \mathbb{N}$. By Exercise 22.5, we conclude that $\mathbb{N} \times \mathbb{N}$ is countable. ∎

Though we did not say so explicitly, $\mathbb{N} \times \mathbb{N}$ is infinite. Therefore, what we have shown is that $\mathbb{N} \times \mathbb{N}$ is equivalent to $\mathbb{N}$.

Exercise 22.9.
Let A be a finite set and let B be a countable set. Prove that $A \times B$ is countable. ○

Corollary 22.10.
If A and B are countable sets, then $A \times B$ is countable.

Assuming you were paying attention to all the previous results, this will not be hard to prove (see Problem 22.8).
We are now ready for the two main theorems of this chapter. After all this work, we can finally show that the set of rational numbers is countably infinite.

Theorem 22.11.
The set of rational numbers, $\mathbb{Q}$, is countably infinite.

What follows is a natural way to attempt to prove this. It is, unfortunately, incorrect. But it's worthwhile to present it, see what goes wrong, and fix it.

Not a proof.
The rationals can be thought of as p/q where p and q are integers with $q \neq 0$. Thus, we can define a map from $\mathbb{Q}$ to $\mathbb{Z} \times (\mathbb{Z} \setminus \{0\})$ by $f(p/q) = (p, q)$. Then f is bijective, so $\mathbb{Q} \approx \mathbb{Z} \times (\mathbb{Z} \setminus \{0\})$. By Corollary 22.10, the latter set is countable. Thus we conclude that $\mathbb{Q}$ is countable. ⍰

As we mentioned (though not quite this dramatically) there's a *HUGE* error in this proof. Find it, fix it, and then read on and see if you really figured it out.

The problem above was that the function was not well-defined. So let's try again. In the proof below, we begin by considering the positive rationals so that we don't have to worry about whether to put the minus sign in the numerator or denominator.

Proof; the real thing.
We will begin by showing that $\mathbb{Q}^+$ is countable. We define $f : \mathbb{Q}^+ \to \mathbb{N} \times \mathbb{N}$ as follows. Write each member of $\mathbb{Q}^+$ as p/q where $p, q > 0$ and p/q is in reduced form; that is, p and q have no positive common factor other than 1. Now define $f(p/q) = (p, q)$. Because p/q is in reduced form, f is well-defined and one-to-one. Since $\mathbb{N} \times \mathbb{N}$ is countable (Theorem 22.8), and $f(\mathbb{Q}^+)$ is a subset of it, we know from Corollary 22.4 that $f(\mathbb{Q}^+)$ is countable. Hence $\mathbb{Q}^+$ is countable. Now the set of negative rationals, $\mathbb{Q}^-$, is equivalent to $\mathbb{Q}^+$. Since $\mathbb{Q} = \mathbb{Q}^+ \cup \mathbb{Q}^- \cup \{0\}$, and we have a finite union of countable sets, we use Corollary 22.7 to conclude that $\mathbb{Q}$ is countable. Since $\mathbb{Q}$ is infinite we know that it is countably infinite. ∎

Looks like we live in a countable world! Not quite—it's time to give an example of an uncountable set. The next theorem will show that the set of real numbers is uncountable.

There's one sticky point in our proof that the reals are uncountable. We will use the decimal representation of real numbers in $(0, 1)$. Thus, a word about decimal expansions is in order here. Each element of $(0, 1)$ has a decimal representation; that is, for $x \in (0, 1)$, there exist integers $a_1, a_2, \ldots, a_n, \ldots$ with $0 \leq a_n \leq 9$ such that $x = 0.a_1 a_2 \ldots a_n \ldots$. In Problem 19.10, we showed that $0.999\ldots = 1.000\ldots$. This means that the number 1 has two representations.

In fact, many numbers in $(0, 1)$ have two decimal representations. It can be shown, however, that the only time this can happen is when the representations are of the form $0.a_1a_2a_3\ldots a_n999\ldots$ for some $n \in \mathbb{Z}^+$, or $0.a_1a_2a_3\ldots a_m1000\ldots$ for some $m \in \mathbb{Z}^+$. When given a choice, we will always choose the representation ending with repeated 9's.

Theorem 22.12.
The set of real numbers, $\mathbb{R}$, is uncountable.

The idea of this proof is due to Georg Cantor and is called Cantor's diagonalization argument.

Proof.
We will suppose, to the contrary, that $\mathbb{R}$ is countable and see what happens. Since every subset of a countable set is countable, the open interval $(0, 1)$ must be countable, too. Since $(0, 1)$ is clearly infinite, and we have shown that $\mathbb{Z}^+$ is countably infinite, there exists a bijective function $f : \mathbb{Z}^+ \to (0, 1)$. We will list the values of f using the decimal expansion of each element of $(0, 1)$. So,

$$
\begin{aligned}
f(1) &= 0.a_{11}a_{12}a_{13}\ldots \\
f(2) &= 0.a_{21}a_{22}a_{23}\ldots \\
f(3) &= 0.a_{31}a_{32}a_{33}\ldots \\
&\vdots
\end{aligned}
$$

where each a_{ij} represents an integer between 0 and 9. Since f is onto, each number in $(0, 1)$ appears in this list.

The odd thing is this: we can construct a number $b = 0.b_1b_2\ldots \in (0, 1)$ not in this list (hence showing that our function cannot possibly be onto) by describing its decimal representation as follows. Look at $f(1)$. If $a_{11} = 2$ let $b_1 = 3$. If, on the other hand, $a_{11} \neq 2$, define $b_1 = 2$. Then the first digits of $f(1)$ and b are different, so b is not $f(1)$. For b_2, if $a_{22} = 2$, let $b_2 = 3$. If, on the other hand, $a_{22} \neq 2$, define $b_2 = 2$. Then the second digits of b and $f(2)$ are different. So b is not $f(2)$. Now compare the element b we have constructed with the list below:

$$
\begin{aligned}
f(1) &= 0.\mathbf{a_{11}}a_{12}a_{13}\ldots \\
f(2) &= 0.a_{21}\mathbf{a_{22}}a_{23}\ldots \\
f(3) &= 0.a_{31}a_{32}\mathbf{a_{33}}\ldots \\
&\vdots \\
f(n) &= 0.a_{n1}a_{n2}a_{n3}\ldots \mathbf{a_{nn}}\ldots
\end{aligned}
$$

$$
b = 0.\mathbf{b_1 b_2 b_3}\ldots
$$

We constructed b so that $b_n \neq a_{nn}$, and therefore $b \neq f(n)$ for every n. Then b can't be in our list, which is a bit bizarre since we claim to have numbered all the elements in $(0, 1)$, and b is certainly one of the things we numbered. This contradiction must mean that we have assumed falsely that $\mathbb{R}$ is countable. ∎

A word of caution: students often forget that some of the theorems proved in previous chapters and some of the definitions only apply to finite sets. Since the notion of finite and infinite is often counter-intuitive, you really must make sure that you check the hypotheses of the theorems you wish to apply *before* you apply the theorems.

Reactions to Cantor's work in set theory were mixed (see, for example, [48, p. 1003]). Leopold Kronecker opposed Cantor's theory and so did Henri Poincaré. In a discussion of Cantor's work, Poincaré [65] said "For my part, and I am not alone, I think that the important thing is never to introduce objects other than those that can be completely defined in a finite number of words."[1] Hilbert and Russell praised Cantor's work; in fact, the first question Hilbert stated in his address to the International Congress of Mathematicians in 1900 had to do with Cantor's set theory (see Spotlight: Hilbert's Seventh Problem, Chapter 27). And, in Hilbert's memorial speech for Hermann Minkowski, Hilbert points out that Minkowski was the first mathematician of their time who understood the importance of Cantor's work. He quotes Minkowski as saying, "History will call Cantor one

---

[1] The translation is ours.

of the most profound mathematicians of our time; it is truly regretful that a very prominent mathematician [here Hilbert tells us that this mathematician is Kronecker] led an opposition not based entirely upon factual grounds, which spoiled Cantor's pleasure in his scientific investigations."[2]

Solutions to Exercises

Solution to Exercise (22.1).

Proof.
Since A and B are countably infinite, we know that $A \approx \mathbb{N}$ and $B \approx \mathbb{N}$. By the transitivity (and symmetry) of the relation $\approx$, we may conclude that $A \approx B$. By the definition of this equivalence relation, there exists a bijective function f mapping A onto B. ∎

Solution to Exercise (22.5).
First suppose that A is countable. If A is finite, then since $A \neq \emptyset$ there exists an integer n and a bijection $f : A \to \{1, 2, \ldots, n\}$. In particular, f is a one-to-one mapping of A into $\mathbb{N}$. So we have found our f, if A is finite. If A is infinite, then A is countably infinite. Therefore, there is a bijection $f : A \to \mathbb{N}$. Thus, in both cases, we have a one-to-one mapping $f : A \to \mathbb{N}$.

Now suppose that we have a one-to-one mapping f of A into $\mathbb{N}$. Then f maps A onto its range. Therefore $A \approx \text{ran}(f)$. But $\text{ran}(f)$ is a subset of $\mathbb{N}$, and by Theorem 22.2 we know that it must be countable. Thus A is countable, as desired.

Solution to Exercise (22.9).
Here's one way to prove this:

---

[2]The translation is ours.

Proof.
If $A = \emptyset$, then $A \times B$ is empty and we are done. Otherwise there exists a positive integer n and a bijective function $f : \{1, 2, \ldots, n\} \to A$. Thus we may write $A \times B$ as a union: $A \times B = \bigcup_{j=1}^{n} (\{f(j)\} \times B)$. It is easy to check that for each j the function $g_j : B \to \{f(j)\} \times B$ defined by $g_j(b) = (f(j), b)$ is a bijection. Therefore, $\{f(j)\} \times B$ is countable for each j. Thus, we have written $A \times B$ as a finite union of countable sets, and by Corollary 22.7 we know that a finite union of countable sets is countable. Thus $A \times B$ is countable. ■

Problems

Problem 22.1.
Give an example, if possible, of each of the following:
  (a) a countably infinite collection of pairwise disjoint finite sets whose union is countably infinite; (See Problem 8.11 for the definition of pairwise disjoint.)
  (b) a countably infinite collection of nonempty sets whose union is finite;
  (c) a countably infinite collection of pairwise disjoint nonempty sets whose union is finite.

Problem 22.2.
Which of the following sets are finite? countably infinite? uncountable? (Be careful—don't apply theorems for finite sets to infinite sets!) Give reasons for your answers for each of the following:
  (a) $\{1/n : n \in \mathbb{Z} \setminus \{0\}\}$;
  (b) $\mathbb{R} \setminus \mathbb{N}$;
  (c) $\{x \in \mathbb{Z} : |x - 7| < |x|\}$;
  (d) $2\mathbb{Z} \times 3\mathbb{Z}$;
  (e) the set of all lines with rational slopes;
  (f) $\mathbb{Q} \setminus \{0\}$;
  (g) $\mathbb{N} \setminus \{1, 3\}$.

Problem 22.3.
Is the set of all infinite sequences of 0's and 1's finite, countably infinite, or uncountable? Guess and then prove, please.

Problem 22.4.
Suppose that $A \subseteq B \subseteq C$, that the sets A and C are equivalent, and that C is countable. Is $A \approx B$? Prove or give a counterexample.

Problem 22.5.
(a) Give an example of two sets A and B, such that $B \subseteq A$ and $B \approx A$, but $B \neq A$.
(b) Prove that if A is a countably infinite set, then there is always a subset B of A such that $B \subset A$ and $B \approx A$.
(c) Prove that if A is an uncountable set, then there is always a subset B of A such that $B \subset A$ and B is also uncountable.
(d) Prove that if A is a finite set, $B \subseteq A$ and $B \approx A$, then $B = A$.

Problem 22.6.
Prove Corollary 22.4.

Problem 22.7.
Prove Corollary 22.7. To do this, note that the corollary can be re-stated a bit more formally as follows. If we have sets $A_1, \ldots, A_n$, and each one is countable, then $\bigcup_{i=1}^{n} A_i$ is countable.

Problem 22.8.
Prove Corollary 22.10.

Problem 22.9.
There is another way to show that $\mathbb{Q}$ is countable. Turn the outline below into a proof by describing the counting process. (Don't try to find a formula for the function.)

Outline of proof.
The proof is simplest if we show that the positive rationals, $\mathbb{Q}^+$, is a countably infinite set. You showed in the exercises (and it is easy to see) that $\mathbb{Q}^- \approx \mathbb{Q}^+$. Then $\mathbb{Q} = \mathbb{Q}^+ \cup \mathbb{Q}^- \cup \{0\}$ is infinite and countable, so $\mathbb{Q} \approx \mathbb{N}$. So we will restrict our attention to $\mathbb{Q}^+$. To see that $\mathbb{Q}^+$ is countable, we will make a chart of all the fractions of the

form m/n where m and n are positive integers; that is, we consider the following array of numbers:

$$
\begin{array}{ccccc}
1 & \frac{2}{1} & \frac{3}{1} & \frac{4}{1} & \cdots \\[2mm]
\frac{1}{2} & \frac{2}{2} & \frac{3}{2} & \frac{4}{2} & \cdots \\[2mm]
\frac{1}{3} & \frac{2}{3} & \frac{3}{3} & \frac{4}{3} & \cdots \\[2mm]
\frac{1}{4} & \frac{2}{4} & \frac{3}{4} & \frac{4}{4} & \cdots \\[2mm]
\cdots & \cdots & \cdots & \cdots & \cdots
\end{array}
$$

Try counting the elements in the array in an orderly fashion. Make sure you don't count numbers twice! ∎

Problem 22.10.

Prove the following generalization of Corollary 22.7.

Theorem 22.13.

If A_i is countable for all $i \in \mathbb{Z}^+$, then $\bigcup_{i \in \mathbb{Z}^+} A_i$ is countable.

Note that induction will not work here. We suggest that you adapt the ideas of the alternate proof of Theorem 22.11 outlined in Problem 22.9.

23

CHAPTER

Metric Spaces

There are many ways to measure distance in the spaces in which we live and work. For example, if you want the shortest distance between two geographical places (the distance "as the crow flies"), you follow the line segment joining them. But in real life this isn't always possible. If you are driving your car through a city or across your campus, you need to go around solid objects and not through them. So how do we calculate distance in those cases? Measuring distance in a set X is a very small (but interesting) part of a branch of mathematics known as "point set topology," and we will look at it in detail in this chapter. We will now often refer to the elements of X as points.

So let's go back to the first time you measured distance. It was probably in $\mathbb{R}$, on a number line, and you learned that the distance between two points x and y was the absolute value of the difference of the two numbers. If we write $d(x, y) = |x - y|$, then d is a function and $d : \mathbb{R} \times \mathbb{R} \to \mathbb{R}$. That's straightforward enough, but now we want to generalize our concept of distance. So let's turn to the essential properties of a distance function.

First, distance shouldn't be negative, so $d(x, y) \geq 0$ for two points x and y, and if the distance satisfies $d(x, y) = 0$, then you didn't move anywhere, so $x = y$. You also surely believe that distance from

x to y should be the distance from y to x. And finally, in Theorem 5.5 (Problem 5.9) you learned the triangle inequality, which said that "if x and y are two real numbers, then $|x+y| \leq |x|+|y|$." In Problem 19.1, you showed how to switch the triangle inequality into a statement about distances. We recall the result of that problem here: For real numbers x, y, and z

$$|x - y| \leq |x - z| + |z - y|.$$

In English, this means that our path will be shorter if we go directly from x to y as opposed to taking a detour through z, which is as it should be. So we would want our general distance function to satisfy something like this too; that is, in our new "d" notation we want $d(x, y) \leq d(x, z) + d(z, y)$ for arbitrary points x, y, and z. So now we will define something that acts like a distance on an arbitrary set X and does all the important things that a distance should do.

Let X be a nonempty set. Then a **metric** on X is a function $d : X \times X \to \mathbb{R}$ satisfying (i)–(iv) below.

(i) (Nonnegativity) For all $x, y \in X$, the function d satisfies $d(x, y) \geq 0$.
(ii) (Definiteness) For all $x, y \in X$, the function d satisfies $d(x, y) = 0$ if and only if $x = y$.
(iii) (Symmetry) For all $x, y \in X$, the function d satisfies $d(x, y) = d(y, x)$.
(iv) (Triangle inequality) For all $x, y, z \in X$, the function d satisfies

$$d(x, y) \leq d(x, z) + d(z, y).$$

A metric is also called a distance function. A set X together with the metric d is called a **metric space** and is denoted by (X, d), or just X when it is clear which distance function we are using. Conditions (i) and (ii) together are usually called **positive definiteness**.

When you learn the definition, don't forget to say "Let X be a nonempty set. Then a metric on X is a function $d : X \times X \to \mathbb{R}\ldots$" These sentences tell us something about d, and cannot be omitted.

In the introduction, we showed that a metric can be defined on $\mathbb{R}$ by $d_u(x, y) = |x - y|$. Though we outlined how to show that d_u is a metric, you should write out the details to complete the proof. This metric is often called the **usual metric** (hence the subscript u)

or the **Euclidean metric** on $\mathbb{R}$, and it is the one upon which your intuition is almost certainly based. A set can have lots of metrics. The next example is a metric on $\mathbb{R}$ that is not the same as the metric given by the absolute value.

Example 23.1.
Define a metric $d_d : \mathbb{R} \times \mathbb{R} \to \mathbb{R}$ by

$$d_d(x, y) = \begin{cases} 0 & \text{if } x = y \\ 1 & \text{if } x \neq y \end{cases}.$$

We will show that d_d is a metric on $\mathbb{R}$. This metric is called the **discrete metric**, and it can really challenge your intuition.

Proof.
It is clear that d_d is a function from $\mathbb{R} \times \mathbb{R} \to \mathbb{R}$. Now let x and y be points of $\mathbb{R}$. We begin with nonnegativity: $d_d(x, y) = 0$ or $d_d(x, y) = 1$, so clearly $d_d(x, y) \geq 0$. Thus, the nonnegativity condition holds. Furthermore, since $d_d(x, y) = 0$ if and only if $x = y$, the definiteness condition holds. For symmetry, note that if $x \neq y$, then $y \neq x$ and consequently $d_d(x, y) = 1 = d_d(y, x)$. If $x = y$, then $d_d(x, y) = 0 = d_d(y, x)$, establishing symmetry. Finally, to establish the triangle inequality, note that if z is a point of $\mathbb{R}$, then we have two cases to consider. In the first case, if $x = y$, then $d_d(x, y) = 0$ and the nonnegativity condition implies that $d_d(x, y) = 0 \leq d_d(x, z) + d_d(z, y)$. In the second case, $x \neq y$, which implies that $z \neq x$ or $z \neq y$ (or both). Therefore, either $d_d(x, z) = 1$ or $d_d(z, y) = 1$ (or both). Thus, $d_d(x, y) = 1 \leq d_d(x, z) + d_d(z, y)$, completing the proof of the triangle inequality. ■

The discrete metric can be defined on every space: the distance between two distinct points is one, and the distance from a point to itself is necessarily zero. The proof that this is a metric on a set X is indistinguishable from the one above. Thus we have an example of a metric on $\mathbb{R}^2$. Example 23.2 and Exercise 23.3 provide us with some other metrics on $\mathbb{R}^2$.

Example 23.2.

On $\mathbb{R}^2$ define a metric by

$$d_u((x_1, x_2), (y_1, y_2)) = \sqrt{(x_1 - y_1)^2 + (x_2 - y_2)^2}.$$

Using Project 27.9, it can be shown that this is actually a metric on $\mathbb{R}^2$. For now, you may accept this fact. This metric is referred to as the **usual metric** or the **Euclidean metric** on $\mathbb{R}^2$. In fact, one may also define the **usual metric on $\mathbb{R}^n$** by

$$d_u((x_1, x_2, \dots, x_n), (y_1, y_2, \dots, y_n)) = \sqrt{\sum_{j=1}^{n} (x_j - y_j)^2}.$$

○

Exercise 23.3.

We now have two examples of metrics on $\mathbb{R}$ and two on $\mathbb{R}^2$. Here are two more metrics on $\mathbb{R}^2$. Before you begin the exercise, familiarize yourself with the metrics by computing various distances. For example, try to find the distance from the point $(1, 3)$ to the points $(-3, 4)$ using the various metrics below.

   (a) Show that $d_{tc}((x_1, x_2), (y_1, y_2)) = |x_1 - y_1| + |x_2 - y_2|$ is a metric on $\mathbb{R}^2$. This metric, d_{tc}, is called the **taxicab** metric. Why would it be called that?

   (b) Show that $d_m((x_1, x_2), (y_1, y_2)) = \max\{|x_1 - y_1|, |x_2 - y_2|\}$ is also a metric on $\mathbb{R}^2$. The metric d_m is sometimes called the **max metric**.

○

The two examples introduced in Exercise 23.3 will appear again in the near future.

A metric tells us when points are close. We studied the notion of "closeness" in Chapter 19 when we studied convergent sequences. You can picture convergence of a sequence to the number L in the following way: a sequence converges to L if for every $\epsilon > 0$, the sequence eventually lies in the open interval $(L - \epsilon, L + \epsilon)$; more precisely, the definition of convergence said, "There exists a real number L such that for every $\epsilon > 0$, there exists a real number N such that $|x_n - L| < \epsilon$ for all $n \geq N$." We return to the idea of finding the limit of a sequence, but this time in a metric space. So given a sequence (x_n) of points in a metric space (X, d), then (as we did before) we say that (x_n) **converges** in X if there exists a point $x \in X$ such that for every $\epsilon > 0$, there exists a real number N such that $d(x_n, x) < \epsilon$ for all $n \geq N$. The value x is called the **limit** of the sequence, we say that the sequence **converges to** x and, as before, we write $x_n \to x$ or $\lim_{n\to\infty} x_n = x$. If the sequence does not converge, we say that it **diverges**. If we consider $X = \mathbb{R}$ with the usual metric, this is exactly the same definition that we had in Chapter 19. Since we allow all sorts of choices for X now, we would like to take this opportunity to point out that the point x must be in the space X—not in some larger space that happens to contain X. If it is clear that x belongs to X, we will often say that the sequence converges, rather than "the sequence converges in X." Also, note that as the metric d changes, the distance between pairs of points changes as well. Therefore, it is conceivable that some sequences will converge in one metric, but not in another.

Exercise 23.4.
Complete the sentences.
(a) Let (x_n) be a sequence in a metric space X with metric d. Let $x \in X$. Then (x_n) does not converge to x if ....
(b) Let (x_n) be a sequence in a metric space (X, d). Then (x_n) does not converge if ....

We'll break tradition and give you the answer to part (a) of the above exercise here, because we need it: A sequence (x_n) does not converge to x if there exists an $\epsilon > 0$ such that for every real number N, there exists $m \in \mathbb{N}$ such that $m \geq N$ and $d(x_m, x) \geq \epsilon$. While an answer to (b) might read "a sequence (x_n) does not converge if for every $x \in X$, the sequence does not converge to x," this will probably not be the most useful formulation of the answer. We leave the more useful version to you. ○

Example 23.5.
We know that $1/n \to 0$ in $\mathbb{R}$ with the usual metric. Show that $(1/n, 1/n) \to (0, 0)$ in $\mathbb{R}^2$ with the usual metric.

Proof.
Let $\epsilon > 0$, and let N be a real number with $N > \sqrt{2}/\epsilon$. If n is an integer with $n \geq N$, then

$$\begin{aligned}
d_u((1/n, 1/n), (0,0)) &= \sqrt{(1/n - 0)^2 + (1/n - 0)^2} \\
&= \sqrt{2}/n \\
&\leq \sqrt{2}/N \qquad\qquad \text{(since } n \geq N) \\
&< \sqrt{2}(\epsilon/\sqrt{2}) \qquad\quad \text{(as } N > \sqrt{2}/\epsilon) \\
&= \epsilon.
\end{aligned}$$

See Figure 23.1 for a graphical illustration of this convergent sequence. ∎

You may wonder where we came up with $\sqrt{2}/\epsilon$. We did it by understanding the problem and devising a plan by working backwards. So what you see here is what happened after we went to a separate sheet of paper, and started with the inequality $\sqrt{2}/n < \epsilon$.

Example 23.6.
In Chapter 19, we showed that the sequence $(1/n)$ converges to 0 in $\mathbb{R}$ with the usual metric. Does $(1/n)$ converge to 0 in the discrete metric?

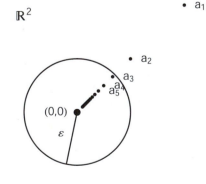

FIGURE 23.1 $(1/n, 1/n) \to (0, 0)$

We claim that the sequence (in $\mathbb{R}$ with the discrete metric) does not converge to 0. To see this, let $\epsilon = 1/2$. For every $N \in \mathbb{R}$, there exists an integer $n \geq N$. Since $1/n \neq 0$, we know that $d_d(1/n, 0) = 1$. Hence for $\epsilon = 1/2$, and for every $N \in \mathbb{R}$, there exists an integer $n \geq N$ such that $x_n = 1/n$ satisfies $d_d(x_n, 0) = d_d(1/n, 0) = 1 \geq 1/2$. Thus $(1/n)$ does not converge to 0. ○

In the discrete metric, every point is "far" from every other point. This makes it very hard to converge.

Exercise 23.7.
Consider $\mathbb{R}$ with the discrete metric. Describe the convergent sequences in this metric space. ○

Sequences have many important properties, some of which we discuss in the problems. The proofs are often quite similar to the proofs we did in Chapter 19. At this point, we give one example of a theorem with such a proof.

Theorem 23.8.
If a sequence (x_n) in a metric space (X, d) converges, then the limit is unique.

The proof of this is the same as the proof of Theorem 19.7.

Solutions to Exercises

Solution to Exercise (23.3).
Parts (a) and (b) are very similar, so we will work part (a) only.
By definition d_{tc} is a function from $\mathbb{R} \times \mathbb{R}$ to $\mathbb{R}$. Now let (x_1, x_2) and (y_1, y_2) be elements of $\mathbb{R}^2$. We will first show the nonnegativity. Because $|a| \geq 0$ for all real numbers a, we know that $d_{tc}((x_1, x_2), (y_1, y_2)) = |x_1 - y_1| + |x_2 - y_2| \geq 0$, showing that nonnegativity of d_{tc} holds. For definiteness, we point out that $d_{tc}((x_1, x_2), (y_1, y_2)) = 0$ if and only if $|x_1 - y_1| + |x_2 - y_2| = 0$. This last equality holds if and only if $|x_1 - y_1| = 0$ and $|x_2 - y_2| = 0$.

This, in turn, holds if and only if $x_1 = y_1$ and $x_2 = y_2$; in other words, $(x_1, x_2) = (y_1, y_2)$. This string of equivalences establishes the definiteness of d_{tc}. Symmetry is shown as follows:

$$d_{tc}((x_1, x_2), (y_1, y_2)) = |x_1 - y_1| + |x_2 - y_2|$$
$$= |y_1 - x_1| + |y_2 - x_2|$$
$$= d_{tc}((y_1, y_2), (x_1, x_2)).$$

To prove that the triangle inequality holds for d_{tc}, let $(z_1, z_2) \in \mathbb{R}^2$. Then

$$d_{tc}((x_1, x_2), (y_1, y_2)) = |x_1 - y_1| + |x_2 - y_2|$$
$$\leq |x_1 - z_1| + |z_1 - y_1| + |x_2 - z_2| + |z_2 - y_2|$$
$$\text{(by the triangle inequality in } \mathbb{R})$$
$$= (|x_1 - z_1| + |x_2 - z_2|) + (|z_1 - y_1| + |z_2 - y_2|)$$
$$= d_{tc}((x_1, x_2), (z_1, z_2)) + d_{tc}((z_1, z_2), (y_1, y_2)).$$

This shows that d_{tc} is a metric on $\mathbb{R}^2$. The taxicab metric between two points measures the distance you have to travel from one point to the next in a city built with rectangular blocks, assuming you stay on the streets, do not take detours, and don't have to worry about one-way streets.

Solution to Exercise (23.4).
The solution to part (a) was given earlier following the exercise, so here is the solution to part (b).

The sequence (x_n) does not converge in the metric space (X, d) if for every $x \in X$ there exists a real number $\epsilon > 0$ such that for every real number N, there exists m such that $m \geq N$ and $d(x_m, x) \geq \epsilon$.

Solution to Exercise (23.7).
We claim that a sequence (x_n) in $(\mathbb{R}, d_d)$ converges if and only if there exist real numbers x and M such that $x_n = x$ for all $n \geq M$. (Such a sequence is called an **eventually constant sequence**.)

First assume that (x_n) is a sequence for which there exist real numbers x and M satisfying $x_n = x$ for all $n \geq M$. We will show that $x_n \to x$. Let $\epsilon > 0$ and let $N = M$. Then for $n \geq N$ we know that $d_d(x_n, x) = d_d(x, x) = 0 < \epsilon$, which shows that (x_n) converges.

For the converse, assume that (x_n) converges. Consider $\epsilon = 1/2$. Then there exists N such that $d_d(x_n, x) < 1/2$ for $n \geq N$. By the definition of d_d, the only way that this can happen is if $x_n = x$ for $n \geq N$. Taking $M = N$, we have shown that there exist x and M such that $x_n = x$ for all $n \geq M$, as desired.

Problems

Unless otherwise specified, assume that you are working in a general metric space (X, d).

Problem 23.1.

(a) Suppose a student writes the following: A metric is a function satisfying (i)–(iv) below.

    (i) (Nonnegativity) $d(x, y) \geq 0$,

    (ii) (Definiteness) $d(x, y) = 0$, if and only if $x = y$,

    (iii) (Symmetry) $d(x, y) = d(y, x)$, and

    (iv) (Triangle inequality) if z is a point in X, then

$$d(x, y) \leq d(x, z) + d(z, y).$$

Write this student a letter indicating what was omitted from the definition, what must be inserted, and what else (if anything) needs to be changed to make it a correct definition.

(b) Suppose the student had exactly the same definition as in the text, except for the triangle inequality, where the student has " $d(x, y) \leq d(x, z) + d(z, y)$ for some $z \in X$."

Write a correct, careful, and complete response to this student.

Problem 23.2.

(a) In $\mathbb{R}$, find the distance of the number 1 to the number 3 in the usual metric and in the discrete metric.

(b) In $\mathbb{R}^2$, find the distance of the point $(1, 3)$ to the point $(2, 5)$ in the usual metric, the taxicab metric, the max metric, and the discrete metric.

Problem 23.3.

(a) Sketch the set $\{(x, y) \in \mathbb{R}^2 : d_u((x, y), (0, 0)) < 1\}$, where d_u is the usual metric.

(b) Sketch the set $\{(x, y) \in \mathbb{R}^2 : d_{tc}((x, y), (0, 0)) < 1\}$, where d_{tc} is the taxicab metric.

(c) Sketch the set $\{(x, y) \in \mathbb{R}^2 : d_m((x, y), (0, 0)) < 1\}$, where d_m is the max metric.

(d) Sketch the set $\{(x, y) \in \mathbb{R}^2 : d_d((x, y), (0, 0)) < 1\}$, where d_d is the discrete metric.

(e) Sketch the set $\{(x, y, z) \in \mathbb{R}^3 : d_u((x, y, z), (0, 0, 0)) < 1\}$, where d_u is the usual metric. (See Example 23.2 for the definition if you need it.)

Problem 23.4.

(a) We defined the max metric on $\mathbb{R}^2$. Define the **max metric** on $\mathbb{R}^n$ and prove that it is a metric.

(b) We defined the taxicab metric on $\mathbb{R}^2$. Define the **taxicab metric** on $\mathbb{R}^n$ and prove that it is a metric.

Problem 23.5.

(a) Show that $d : \mathbb{R} \times \mathbb{R} \to \mathbb{R}$ defined by

$$d((x_1, x_2), (y_1, y_2)) = |x_1 - y_1|$$

is not a metric on $\mathbb{R}^2$.

(b) Is $d : \mathbb{R} \times \mathbb{R} \to \mathbb{R}$ defined by

$$d((x_1, x_2), (y_1, y_2)) = (x_1 - y_1)^2 + (x_2 - y_2)^2$$

a metric on $\mathbb{R}^2$?

Problem 23.6.

Let (X, d) be a metric space. Let α be a real number and define a new function d_α on $X \times X$ by $d_\alpha(x, y) = \alpha d(x, y)$. Is d_α a metric on X? If not, what assumptions must be placed on α to assure that d_α is a metric? Prove your answer.

Problem 23.7.

A set F in a metric space (X, d) is **bounded** if there exists a positive number M such that $d(x, y) \leq M$ for all $x, y \in F$.

(a) Consider the following "not a definition" of a bounded set.
"A set is bounded if for each $x, y \in F$ there exists a positive number M such that $d(x, y) \le M$."
Give a complete, clear, concise explanation of the problems with this definition.

(b) Give an example of a metric space and an infinite set that is bounded in that metric. Prove that it is bounded.

(c) Complete the following definition: Let X be a metric space with metric d and let F be a subset of X. Then F is not bounded if
. . . .

(d) Give an example of a metric space and a set that is not bounded in that metric. Prove that it is not bounded.

Problem 23.8.

Let X be a set with a metric d. Define a function $d_b : X \times X \to \mathbb{R}$ by

$$d_b(x, y) = \min\{d(x, y), 1\}.$$

(a) Show that d_b is a metric on X. This metric is called the **bounded metric associated with** d on X.

(b) (This part uses Problem 23.7.) Consider the metric space (X, d_b). Show that in this space, every subset of X is bounded.

Problem 23.9.

Show that in a metric space (X, d) the metric satisfies

$$|d(x, z) - d(y, z)| \le d(x, y),$$

for all $x, y, z \in X$.

Problem 23.10.

Let X be the space of polynomials with real coefficients. Define a function d from $X \times X \to \mathbb{R}$ by $d(p, q) = |p(0) - q(0)|$. Is d a metric? If so, prove it. If not, why not?

Problem 23.11.

The following exercise is only appropriate if you have had integration in calculus.

Let X be the space of real-valued continuous functions defined on the interval $[0, 1]$. Define a function $d : X \times X \to \mathbb{R}$ by

$$d(f, g) = \int_0^1 |f(t) - g(t)| dt,$$

for all $f, g \in X$.
  (a) Show that d is a metric.
  (b) Find the distance between e^x and $\sin(\pi x/2)$.

Problem 23.12.
Choose a fixed point x_0 in $\mathbb{R}^2$. If d_u denotes the usual (or Euclidean) metric on $\mathbb{R}^2$, then we define $d : \mathbb{R}^2 \times \mathbb{R}^2 \to \mathbb{R}$ by

$$d(x, y) = \begin{cases} d_u(x, y) & \text{if } x \text{ and } y \text{ are on a straight} \\ & \text{line through } x_0 \\ d_u(x, x_0) + d_u(x_0, y) & \text{otherwise.} \end{cases}$$

Figure 23.2 illustrates this function for three pairs of points in the plane. Prove that d is a metric on $\mathbb{R}^2$.

This metric is sometimes called the "French railway system metric." (See [39, p. 56].) Why? Think of x_0 as Paris, and you'll note that all trains pass through Paris, whether they need to or not.

Problem 23.13.
Prove each of the following.
  (a) Consider $\mathbb{R}^2$ with the max metric. Prove that $(1/n, 2/n) \to (0, 0)$.
  (b) Consider $\mathbb{R}$ with the usual metric. Prove that $(-1)^n n/(3n+1) \not\to 0$.

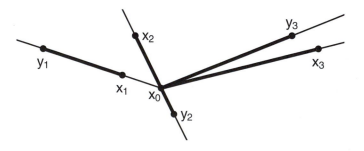

FIGURE 23.2 The metric: $d(x_i, y_i)$ for $i = 1, 2, 3$

(c) Consider $\mathbb{R}^2$ with the max metric. Does $((-1)^n, 2/n))$ converge in this space?

Problem 23.14.

Consider $\mathbb{Z}$ with the usual metric.
(a) Show that a sequence that is eventually constant converges; that is, if there exists an integer m such that $x_n = x_k$ for all $n, k \geq m$, then the sequence converges.
(b) Can you give other examples of convergent sequences in $(\mathbb{Z}, d_u)$? Explain your answer.

Problem 23.15.

Let (X, d) be a metric space, and let (x_n) be a convergent sequence in X.
(a) Prove that there exists $x \in X$ and a natural number K such that

$$d(x_n, x) \leq K \text{ for all } n \in \mathbb{N}.$$

(You should know a similar problem.)
(b) Prove that the set $\{x_n : n \in \mathbb{N}\}$ is bounded, as defined in Problem 23.7; that is, prove that there exists a positive number M such that $d(x_n, x_m) \leq M$ for all $n, m \in \mathbb{N}$.

Problem 23.16.

In Problem 19.14, part (c), we defined the term Cauchy sequence and proved some facts about such sequences. This problem asks you to do the same in a general metric space.
(a) Define Cauchy sequence in a metric space (X, d).
(b) Prove that if (x_n) converges in (X, d), then (x_n) is Cauchy.

Problem 23.17.

(This problem uses Problem 23.16.) Let $X = \mathbb{R} \setminus \mathbb{Q}$ with the usual metric d_u. Prove that the sequence (x_n), where $x_n = \sqrt{2}/n$, is a Cauchy sequence in X, but (x_n) does not converge in X.

24

CHAPTER

Getting to Know Open and Closed Sets

When we work in $\mathbb{R}$ with the usual metric, we think of distance as measured by absolute value. Points are close when the absolute value of the difference is small. In other words, we might reasonably argue that points x and y are close when they satisfy $|y - x| < r$, where r is a small positive number; that is to say, y is in the open interval $(x - r, x + r)$. This interpretation allows us to visualize the distance between the points. As it turns out, all metrics have this visual interpretation.

Let x be a point in a metric space (X, d) and let r be a real number with $r > 0$. Then the **open ball of radius** r **about** x is denoted $B_d(x, r)$ and is defined by $B_d(x, r) = \{y \in X : d(y, x) < r\}$. We will call $B_d(x, 1)$ the **open unit ball about** x. Note that the radius of the open ball $B_d(x, r)$ is always positive, and $B_d(x, r)$ is centered at x. This is quite an important definition, and we will be able to do a lot with it. But remember, before you work an example, state a theorem or write a proof, make sure that you and your intended reader are clear on the space you are working on, the metric you are using, and what you want to show.

Example 24.1.
Consider the set $\mathbb{R}$ with the usual metric. What does $B_{d_u}(1, 1/2)$ mean? What is $B_{d_u}(x, r)$, for an arbitrary $x \in X$ and $r > 0$?

By definition, $B_{d_u}(1, 1/2) = \{y \in \mathbb{R} : d_u(y, 1) < 1/2\}$. The notation is preventing us from seeing something we all know pretty well, so let's get rid of it. Rewriting,

$$
\begin{aligned}
B_{d_u}(1, 1/2) &= \{y \in \mathbb{R} : |y - 1| < 1/2\} \\
&= \{y \in \mathbb{R} : -1/2 < y - 1 < 1/2\} \\
&= \{y \in \mathbb{R} : 1/2 < y < 3/2\} \\
&= (1/2, 3/2).
\end{aligned}
$$

Figure 24.1 shows $B_{d_u}(1, 1/2)$ graphically.

Now the solution of the general case is the same (there's also a sketch of what's happening in Figure 24.2): By definition, $B_{d_u}(x, r) = \{y \in \mathbb{R} : d_u(y, x) < r\} = \{y \in \mathbb{R} : |y - x| < r\} = \{y \in \mathbb{R} : -r < y - x < r\} = \{y \in \mathbb{R} : x - r < y < x + r\} = (x - r, x + r)$. Therefore, $B_{d_u}(x, r) = (x - r, x + r)$. ○

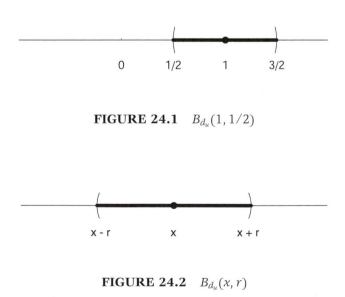

| 0 | 1/2 | 1 | 3/2 |

FIGURE 24.1 $B_{d_u}(1, 1/2)$

| x - r | x | x + r |

FIGURE 24.2 $B_{d_u}(x, r)$

The next example shows that the open balls depend on what the underlying set X is.

Example 24.2.
Consider the set $X = [0, 1)$ with the usual metric. What is $B_{d_u}(1/4, 2/3)$?

Since the center is $x = 1/4$ and the radius is $r = 2/3$, we find

$$
\begin{aligned}
B_{d_u}(1/4, 2/3) &= \{x \in [0, 1) : |x - 1/4| < 2/3\} \\
&= \{x \in [0, 1) : -2/3 < x - 1/4 < 2/3\} \\
&= \{x \in [0, 1) : -5/12 < x < 11/12\} \\
&= [0, 11/12).
\end{aligned}
$$
○

Now it's your turn.

Exercise 24.3.
Consider the set $\mathbb{R}^2$ with the usual metric (defined in Example 23.2). What is the set $B_{d_u}((0, 1), 4)$? Describe the set using precise set notation and sketch it. ○

These balls can be used to describe the basic structure of metric spaces. For example, for two distinct points x and y in a metric space, we can always find two disjoint open balls, B_x and B_y, such that $x \in B_x$ and $y \in B_y$. This probably agrees with your intuition. On the other hand, as we shall see, there exist metric spaces in which sets consisting of a single point are open balls! In the remainder of this chapter, we will see how these balls can be used to determine which sequences converge. But before we can develop the connection between balls and convergence, we need to look at some of the important sets we can make using these balls together with the set operations we studied earlier.

In a metric space (X, d), a subset U of X is **open** if for every point $x \in U$, there exists an open ball $B_d(x, r)$ satisfying $B_d(x, r) \subseteq U$. Note that r depends on x, so that as x changes, r will too. This definition is very visual, so pictures will help you. A word of caution is in order before we begin our examples: Throw away any preconceived notions you have about what an open ball should look like.

Example 24.4.
Show that the interval $(2, 4)$ is open in $\mathbb{R}$ with the usual metric.

You may be thinking that there's nothing to show; after all, it's an open interval. But we still need to use the definition to check that this interval is open in the sense we have defined here, so let's briefly review exactly what we have to show.

We will show that *for every* $x \in (2, 4)$, there is an open ball $B_{d_u}(x, r) \subseteq (2, 4)$. We know our metric is the usual one, so by Example 24.1, we know we need to show that there exists a positive number r with $(x - r, x + r) \subseteq (2, 4)$. To find r, you should draw pictures whenever you can. Before you read our solution, find your own following this outline: first, draw and label an appropriate picture, which you will then use as you continue on in this proof. Next, pick an *arbitrary* point $x \in (2, 4)$. Now, we need to find a positive number r with $B_{d_u}(x, r) = (x - r, x + r) \subseteq (2, 4)$. Look at your picture to find a possible value of r. Then show that it works.

Proof.
We will prove that $(2, 4)$ is open. Let $x \in (2, 4)$. Then $2 < x < 4$, so both $x - 2$ and $4 - x$ are positive. Let $r = \min\{x - 2, 4 - x\}$. (See Figure 24.3.) Then $r > 0$. Now we'll check that $B_{d_u}(x, r) \subseteq (2, 4)$. So let $y \in B_{d_u}(x, r)$. Then, by definition, $|y - x| < r$. Therefore, $-r < y - x < r$. From the upper inequality we obtain $y - x < 4 - x$, and hence $y < 4$. From the lower inequality we obtain $-(x - 2) < y - x$, and hence $2 < y$. Thus, $y \in (2, 4)$, and we conclude that $B_{d_u}(x, r) \subseteq (2, 4)$. ∎

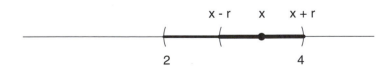

FIGURE 24.3 $B_{d_u}(x, r) \subseteq (2, 4)$

A few questions and comments are in order. First, don't forget to check that $r > 0$. An open ball of 0 radius, or (worse yet) negative radius, is no ball at all. Choosing r as the minimum of finitely many real numbers is a pretty standard thing to do, so it's a good idea to get used to it now. And finally, there are lots of choices for r. We picked one value that worked. Every smaller positive value would work too.

Exercise 24.5.
Let (X, d) be a metric space and let $U \subseteq X$. Complete the sentence: The set U is not open in (X, d) if ....

This is one of those exercises where you will want to check your solution against ours before you go on. So here's ours: The set U is not open in (X, d) if there exists a point $x \in U$ such that for every open ball $B_d(x, r)$ about x, there exists a point $y \in B_d(x, r) \cap U^c$.   ○

Exercise 24.6.
Show that the interval $[2, 4]$ is not open in $\mathbb{R}$ with the usual metric.
                                                                                              ○

Exercise 24.7.
The setting: $\mathbb{R}^2$ with the usual metric. Your mission: to show that the open unit ball, $B_{d_u}((0, 0), 1)$, about $(0, 0)$ is open.

Let the following steps guide you:

(a) Sketch the open ball of radius 1 about the point $(0, 0)$.
(b) Without thinking too much about it, choose a point in the open ball (make sure you don't choose $(0, 0)$). Make a dot at the point, and label it (a, b).
(c) Draw as large an open ball as you can that is still contained in $B_{d_u}((0, 0), 1)$ and centered at your dot (a, b). What's the radius of that open ball? (Here's a potentially helpful suggestion: draw the radius of the open ball $B_{d_u}((0, 0), 1)$ that passes through the point (a, b).)

(d) Now you are ready to start the problem. Find a positive real number r such that $B_{d_u}((a, b), r)$ appears to be contained in $B_{d_u}((0, 0), 1)$.

(e) Show that $B_{d_u}((a, b), r) \subseteq B_{d_u}((0, 0), 1)$. Write out the whole proof carefully. Include your picture; it's very helpful for the writer and the reader. ○

There are lots of interesting sets in a metric space, all building on the notion of open ball. We have already introduced open sets. We now come to closed sets. A set E in a metric space X is **closed** if and only if the complement, E^c, is open. So, since the complement of an arbitrary open set is a closed set, we can immediately write down several closed sets. For example, in $\mathbb{R}$ with the usual metric, the set $(-\infty, 2] \cup [4, \infty)$ must be closed, since its complement is the open set $(2, 4)$. (See Example 24.4).

It's important to know many ways to show that sets are open, closed, or neither. Here's a useful result that should have a one-line proof.

Theorem 24.8.
Let (X, d) be a metric space. A subset U of X is open if and only if its complement is closed.

Exercise 24.9.
Find the one- (or two-) line proof of Theorem 24.8. ○

We now have examples of open sets and of closed sets. But, as you will see as you work the next two exercises, things often get a bit complicated.

Exercise 24.10.
Give an example of a set that is neither open nor closed in $\mathbb{R}^2$ with the usual metric. ○

Exercise 24.11.
Let (X, d) be a metric space. Is the empty set open? closed? both? neither? ○

You might have found yourself concluding that if a set is not open, then it is closed. This is normal, because in ordinary English if a door is not open, then it is closed. Unfortunately, in mathematics, that's false! In the two exercises above, we have seen examples of sets that are neither open nor closed, and examples of sets that are both open and closed. Don't assume anything when you work the problems: if we didn't prove it, state it, or use it, then it may not be true.

We defined an open ball and an open set. You will show (in Problem 24.13) that every open ball is an open set, but since this is so important, we'll state it as a theorem. It's interesting to note that the proof is very much like the proof of Exercise 24.7.

Theorem 24.12.
Let (X, d) be a metric space. For every point $x \in X$, and every positive real number r, the set $B_d(x, r)$ is open.

Now we will get to see some ways that we can use open sets and some more odd properties of metric spaces. The first theorem tightens the relationship between open sets and open balls.

Theorem 24.13.
Let (X, d) be a metric space. A set U is open if and only if there is a subset I of X and a set of radii $\{r_y \in \mathbb{R}^+ : y \in I\}$ such that $U = \bigcup_{y \in I} B_d(y, r_y)$.

There are some things that might be confusing to you in this statement, but it's much easier to see what it means to be an open set if you understand Theorem 24.13. The index set is a way of saying that we don't know how many y we have; there could be finitely many or not, countably many or not, and this way we don't have to deal with that issue. Next, the r_y might confuse you. Each ball has a (positive) radius, and if we wrote $B_d(y, r)$ for all y, we would be saying that all the balls have the same radius, r. That's not what the theorem says, so we shouldn't say that either. By using the notation r_y, we allow each y in I to have its own radius, r_y. Having said all this, we now begin the proof.

Proof.
Suppose first that there is a subset I of X such that

$$U = \bigcup_{y \in I} B_d(y, r_y).$$

By the definition of open set, we need to show that for an arbitrary $x \in U$, there exists an open ball $B_d(x, r_x)$ contained in U. Now if $x \in U$, then there exists an element $z \in I$ such that $x \in B_d(z, r_z)$. By Theorem 24.12, the ball $B_d(z, r_z)$ is an open set, and therefore there exists a positive real number r_x such that $B_d(x, r_x) \subseteq B_d(z, r_z)$. Since $B_d(z, r_z) \subseteq U$, we know that $B_d(x, r_x) \subseteq U$. Hence the set U is open.

Now suppose that U is open. We have to find a collection of open balls such that U is the union of those open balls. By the definition of open set, if $x \in U$, there exists an open ball $B_d(x, r_x)$ with $B_d(x, r_x) \subseteq U$. Now we claim that $U = \bigcup_{x \in U} B_d(x, r_x)$. If we establish this claim, our proof will be complete. To see that U is contained in the union, note that if $y \in U$, then $y \in B_d(y, r_y)$, and therefore $y \in \bigcup_{x \in U} B_d(x, r_x)$. Thus, $U \subseteq \bigcup_{x \in U} B_d(x, r_x)$. To show that U contains the union, note that $B_d(x, r_x) \subseteq U$ for each x. From this it is easy to see[1] that $\bigcup_{x \in U} B_d(x, r_x) \subseteq U$, completing the proof. ∎

Theorem 24.13 can be restated as follows: A set U in a metric space X is open if and only if U is a union of open balls.

The proofs of many of the theorems in this chapter provide an excellent opportunity for you to apply all the techniques that you have learned in this course. For this reason, we have left many as problems. Here's another useful theorem.

Theorem 24.14.
An arbitrary union of open sets is open.

The proof of this is left as a problem (Problem 24.11) for you, the reader. By "arbitrary union" we mean that we don't know how many sets we have. So make sure that you don't accidentally assume that there are finitely many sets, or even countably many.

---

[1] If this isn't easy to see, show it using an element-chasing argument. In fact, when you worked Exercise 8.10, you already showed it.

Theorem 24.15.

An arbitrary intersection of closed sets is closed.

The proof of this is left for you to do (Problem 24.12). If you have been paying close attention to the theorems and definitions presented thus far, this should follow from Theorem 24.14. What about an intersection of open sets? a union of closed sets? The results are given below and the proofs are outlined in the problems.

Theorem 24.16.

Let $U_1, \ldots, U_n$ be open sets. Then $\bigcap_{j=1}^{n} U_j$ is an open set.

Theorem 24.17.

Let $F_1, \ldots, F_n$ be closed sets. Then $\bigcup_{j=1}^{n} F_j$ is a closed set.

We'll conclude this chapter with the metric we promised would challenge your intuition.

Example 24.18.

Consider $\mathbb{R}$ with the discrete metric, $d_{\bar{d}}$. Prove the following.
  (a) For each point $x \in \mathbb{R}$, the set $\{x\}$ is an open ball.
  (b) Every set in $(\mathbb{R}, d_{\bar{d}})$ is open.
  (c) Every set in $(\mathbb{R}, d_{\bar{d}})$ is closed.

For part (a), note that for $x \in \mathbb{R}$, the set $\{x\} = B_{d_{\bar{d}}}(x, 1/2)$. By Theorem 24.12, the set $B_{d_{\bar{d}}}(x, r)$ is an open set and, consequently, $\{x\}$ is open.

For part (b), let S be a subset of $\mathbb{R}$. Since $S = \bigcup_{s \in S}\{s\}$, from part (a) we see that S is a union of open sets. By Theorem 24.14, S is open.

For part (c), let T be a subset of $\mathbb{R}$. Then T^c is also a subset of $\mathbb{R}$, and it follows from part (b) that T^c is open. But a set is closed if and only if its complement is open, and therefore T is closed. ○

There's lots more that we can do here, and we will do it in the problems.

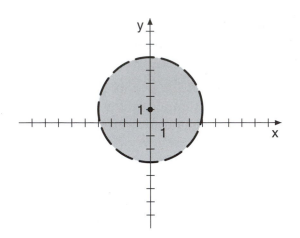

FIGURE 24.4　$B_{d_u}((0, 1), 4)$.

Solutions to Exercises

Solution to Exercise (24.3).

We first calculate $B_{d_u}((0, 1), 4)$. The graphical representation is shown in Figure 24.4.

$$
\begin{aligned}
B_{d_u}((0, 1), 4) &= \{(x, y) \in \mathbb{R}^2 : d_u((x, y), (0, 1)) < 4\} \\
&= \{(x, y) \in \mathbb{R}^2 : \sqrt{x^2 + (y - 1)^2} < 4\} \\
&= \{(x, y) \in \mathbb{R}^2 : x^2 + (y - 1)^2 < 16\}.
\end{aligned}
$$

Solution to Exercise (24.6).

We will use the solution to Exercise 24.5: Choose $x = 2$ and note that $2 \in [2, 4]$. Let $B_{d_u}(2, r)$ be an open ball about 2. Then r is a positive real number. We claim that $2 - r/2 \in B_{d_u}(2, r) \cap [2, 4]^c$. Since $d_u(2 - r/2, 2) = |(2 - r/2) - 2| = r/2 < r$, we conclude that $2 - r/2 \in B_{d_u}(2, r)$. Also, since $2 - r/2 < 2$ we know that $2 - r/2 \notin [2, 4]$. This establishes the claim and we have shown that $[2, 4]$ is not open in $\mathbb{R}$ with the usual metric.

Solution to Exercise (24.7).

We follow the outline provided beginning with the illustration in Figure 24.5.

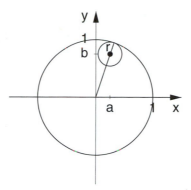

FIGURE 24.5 Sketch to find r, the radius of the open ball centered at (a, b).

Let $(a, b) \in B_{d_u}((0, 0), 1)$. We claim that if we let $r = 1 - \sqrt{a^2 + b^2}$, then $B_{d_u}((a, b), r)$ is an open ball about (a, b) satisfying $B_{d_u}((a, b), r) \subseteq B_{d_u}((0, 0), 1)$.

First note that since $(a, b) \in B_{d_u}((0, 0), 1)$, we have $d_u((a, b), (0, 0)) = \sqrt{a^2 + b^2} < 1$. Hence $r = 1 - \sqrt{a^2 + b^2} > 0$ and $B_{d_u}((a, b), r)$ is an open ball about (a, b).

To prove the set inclusion, let $(x, y) \in B_{d_u}((a, b), r)$. Then

$$d_u((x, y), (0, 0)) \leq d_u((x, y), (a, b)) + d_u((a, b), (0, 0))$$

$$\text{(by the triangle inequality for distances)}$$

$$< r + \sqrt{a^2 + b^2} \quad \text{(since } (x, y) \in B_{d_u}((a, b), r))$$

$$= 1 - \sqrt{a^2 + b^2} + \sqrt{a^2 + b^2} = 1.$$

Hence $(x, y) \in B_{d_u}((0, 0), 1)$. This establishes the second part of the claim.

The definition of an open set implies that $B_{d_u}((0, 0), 1)$ is open.

Solution to Exercise (24.9).
By the definition of closed set, the set U^c is closed in a metric space (X, d) if and only if $(U^c)^c = U$ is open in X.

Solution to Exercise (24.10).
The set $\{(x, y) : 0 < x \leq 1\}$ is neither open nor closed in $\mathbb{R}^2$. We leave the rigorous proof of this to you.

Solution to Exercise (24.11).
The empty set is open in every metric space (X, d). The reason is that the antecedent "$x \in \emptyset$" is always false. This means that the defining implication is always true for $\emptyset$.

The whole space X is also always open: For all $x \in X$, we know that $B_d(x, r) \subseteq X$ for every positive real r. Since $\emptyset^c = X$ and X is open, the definition of closed implies that $\emptyset$ is closed.

Thus we have two examples of sets that are both open and closed: X and $\emptyset$.

Problems

We continue to assume that X is a metric space with metric d unless otherwise stated.

Problem 24.1.
Show that the set $(1, 3) \cup (4, 5)$ is open in $\mathbb{R}$ with the usual metric.

Problem 24.2.
Consider $\mathbb{R}$ with the usual metric. In each case, give an example of a nonempty closed set C and a nonempty open set U such that
  (a) $U \cap C$ is open.
  (b) $U \cup C$ is closed.
  (c) $U \cap C$ is neither open nor closed.
  (d) $U \cup C$ is neither open nor closed.

Problem 24.3.
Prove that in $\mathbb{R}^2$ with the max metric, d_m, the set $\{(x, y) \in \mathbb{R}^2 : -1 < x < 0, -1 < y < 1\}$ is an open set. Include a sketch with your proof.

Problem 24.4.
Decide whether the statements below are true or false. If the statement is true, give a brief reason why. If the statement is false, give a counterexample.
  (a) In $\mathbb{R}$ with the usual metric, the interval $[0, \infty)$ is a closed set.

(b) In $\mathbb{R}$ with the discrete metric, the interval $[0, \infty)$ is an open set.
(c) A finite union of open sets is an open set.
(d) An arbitrary union of closed sets is a closed set.

Problem 24.5.

Consider the set $\mathbb{R}^+$ with the usual metric.
(a) Show that the set $(1, \infty)$ is an open set in $(\mathbb{R}^+, d_u)$.
(b) Show that the set $(0, 1]$ is a closed set in $(\mathbb{R}^+, d_u)$.

Problem 24.6.

Let (X, d) be a metric space.
(a) Prove that for all x and y in X, if $y \neq x$, then there exists an open ball centered at y, say $B_d(y, r)$, such that $x \notin B_d(y, r)$.
(b) Prove that if $x \in X$, then $\{x\}$ is a closed set.

Problem 24.7.

Let (X, d) be a metric space. Let U_j be a sequence of open sets.
(a) Give an example to show that $\bigcap_{j=1}^{\infty} U_j$ may not be open.
(b) Is it ever true that $\bigcap_{j=1}^{\infty} U_j$ is open?

Problem 24.8.

Show that in $\mathbb{R}^2$ with the usual metric, the set $E = \{(x, y) \in \mathbb{R}^2 : 1 < x^2 + y^2 < 4\}$ is open.

Problem 24.9.

Complete the following definition. Let X be a metric space with metric d and let F be a subset of X. Then F is not closed if ....

Problem 24.10.

(a) Give three examples of sets that are both open and closed in $\mathbb{R}$ with the discrete metric.
(b) Give two examples of sets that are both open and closed in $\mathbb{R}$ with the usual metric.
(c) Give an example of a set that is neither open nor closed in $\mathbb{R}^2$ with the max metric. Prove that it is neither open nor closed!
(d) Give an example of a set that is closed and not open in $\mathbb{R}^2$ with the usual metric. Prove that it is closed and not open!

Problem 24.11.
Prove Theorem 24.14. (In other words, show that if $\{O_\alpha : \alpha \in I\}$ is a collection of open sets, then $\bigcup_{\alpha \in I} O_\alpha$ is open.)

Problem 24.12.
Prove Theorem 24.15.

Problem 24.13.
Prove Theorem 24.12.

Problem 24.14.
Let (X, d) be a metric space, $x, y \in X$, and let r_1 and r_2 be positive real numbers with $r_1 < r_2$. Do not use Theorem 24.16.
 (a) Show that $B_d(x, r_1) \subseteq B_d(x, r_2)$.
 (b) From the previous problem we know that every open ball is open. Show that if $B_d(x, r_1)$ and $B_d(y, r_2)$ are open balls, then $B_d(x, r_1) \cap B_d(y, r_2)$ is an open set. Is it an open ball? (Justify your answer to this last question, please.)

Problem 24.15.
Prove Theorem 24.16 by completing both steps below. You may find it very helpful to work Problem 24.14 first.
 (a) Show that the intersection of two open sets is an open set.
 (b) Show that the intersection of finitely many open sets is an open set.

Problem 24.16.
Prove Theorem 24.17. If you did Problem 24.15, you might consider using that result here.

Problem 24.17.
 (a) Let x and y be two distinct points in X and let $r = d(x, y)/2$. Show that $B_d(x, r)$ and $B_d(y, r)$ are disjoint sets.
 (b) Show that for two distinct points x and y in a metric space, there exist disjoint open sets $\mathcal{O}_x$ and $\mathcal{O}_y$ with $x \in \mathcal{O}_x$ and $y \in \mathcal{O}_y$.

Problem 24.18.
Let (X, d_X) and (Y, d_Y) be metric spaces. A function $f : (X, d_X) \to (Y, d_Y)$ **preserves distances** if $d_Y(f(x), f(x')) = d_X(x, x')$ for all x, x' in X.
  (a) In $\mathbb{R}$ with the usual metric, the function $f : (\mathbb{R}, d_u) \to (\mathbb{R}, d_u)$ defined by $f(x) = x$ obviously preserves distances. Give an example of another function that preserves distances.
  (b) Is every function that preserves distances one-to-one? Either prove this statement or give a counterexample.

Problem 24.19.
Let E be a subset of a set X with metric d. A point x is said to be an **interior point** of E if there exists an open ball $B_d(x, r)$ with $B_d(x, r) \subseteq E$. The set of all interior points is called **the interior** of E and is denoted by E^o.
  (a) In $\mathbb{R}$ with the usual metric, and $E = (2, 4]$, show that $4 \notin E^o$. Then show that $(2, 4) = E^o$.
  (b) In $\mathbb{R}^2$ with the max metric, find E^o if $E = \{(x, y) : |x| \le 1\}$.

Problem 24.20.
This problem is only appropriate if you completed Problem 24.19. Let (X, d) be a metric space, and E be a subset of X.
  (a) By the definition of interior point, if $x \in E^o$, then there exists an open ball $B_d(x, r)$ centered at x such that $B_d(x, r) \subseteq E$. Show that, in fact, for each point $x \in E^o$, there exists an open ball $B_d(x, r_x) \subseteq E^o$. Use this to prove that E^o is an open set.
  (b) Prove that a set E is open if and only if every point of E is an interior point. Conclude that a set E is open if and only if $E = E^o$.

Problem 24.21.
Let (X, d) be a metric space. Let $E \subseteq X$. A point $x \in X$ is a **limit point** of E if every open set containing x contains a point $y \in E$ with $y \ne x$. Let E_l denote the set of all limit points of the set E.
  (a) Complete the following definition. A point $x \in X$ is not a limit point of E if . . . .
  (b) What are the limit points of the interval $(2, 4]$ in $\mathbb{R}$ with the usual metric? the discrete metric?

(c) What are the limit points of $B_{d_u}((0,0),1)$ in $\mathbb{R}^2$ with the usual metric? the discrete metric?

(d) What are the limit points of $\{1/n : n \in \mathbb{Z}^+\}$ in $\mathbb{R}$ with the usual metric? the discrete metric?

Problem 24.22.

(This problem assumes that you have completed Problem 24.21.) Let (X,d) be a metric space. Let E be a subset of X. Show that x is a limit point of a set E if and only if every open ball $B_d(x,r)$ about x contains a point $y \in E$ with $y \neq x$.

Problem 24.23.

(This problem assumes that you have completed Problem 24.21.) Let (X,d) be a metric space. Let E be a subset of X.

(a) Prove that x is a limit point of a set E if and only if every open set about x contains infinitely many points different from x.

(b) Prove that a finite set has no limit points.

Problem 24.24.

(This problem assumes that you have completed Problem 24.21.) Let (X,d) be a metric space. Let E be a subset of X.

(a) Show that E is closed if and only if E contains all its limit points. In other words, prove that E is closed if and only if $E_l \subseteq E$.

(b) Let E be a set. The **closure** of E is denoted by $\overline{E}$ and is defined by $\overline{E} = E \cup E_l$. Show that if x is a limit point of $\overline{E}$, then x is a limit point of E.

(c) Show that if x is a limit point of $\overline{E}$, then $x \in \overline{E}$ (note that you did the hard part in (b) above). Conclude that $\overline{E}$ is closed.

Problem 24.25.

Assuming that X is a metric space with metric d, define a new function $d_e : X \times X \to \mathbb{R}$ by

$$d_e(x,y) = \frac{d(x,y)}{1 + d(x,y)}.$$

Show that d_e is also a metric on X.

25

Modular Arithmetic

CHAPTER

You are quite familiar with adding, subtracting, multiplying, and dividing integers, rational numbers, real numbers, and complex numbers. But this is not the only kind of arithmetic to be found within mathematics. Although you may have never seen it explained with mathematical precision, there's a very different kind of arithmetic that you use every single day. The popular name for these calculations is *clock arithmetic*, and it is indeed based upon the clock.

Consider the following scenario: Suppose it is now 3:00 P.M., and you start on a 28-hour trip. What time will it be when you return? A quick calculation yields an answer of 7:00 P.M. How did you arrive at this answer? You did something we all find natural—you did clock arithmetic. We will now carefully build upon this idea, and we will apply many of the concepts we have already covered to help us understand it.

Clock arithmetic isn't done on numbers, but rather on equivalence classes of numbers. So we first need to find the right equivalence relation. Recall that for two integers a and b with $a \neq 0$, we say that a divides b, written $a \mid b$, if there is an integer k such that $b = ak$. Now we are ready for the equivalence relation. Let $n \in \mathbb{Z}$ such that $n > 1$. Two integers x and y will be related if $n|(y - x)$. In

313

this case we say x **is congruent to** y **modulo** n, and we will write $x \equiv y \pmod{n}$.

Exercise 25.1.
Find five different solutions to each of the problems below, and then find five integers that are not solutions.
  (a) Find $x \in \mathbb{Z}$ such that $5 \equiv x \pmod{12}$.
  (b) Find $x \in \mathbb{Z}$ such that $x > 1$ and $-3 \equiv 39 \pmod{x}$. ○

Theorem 25.2.
Let $n > 1$ be an integer. The relation congruence modulo n is an equivalence relation on $\mathbb{Z}$.

The proof of this theorem is left as Problem 25.3.

For an integer $n > 1$, the set of all equivalence classes with respect to the relation congruence modulo n is called the **integers modulo** n and denoted by $\mathbb{Z}_n$. It follows from Theorem 11.4 that $\mathbb{Z}_n$ is a partition of $\mathbb{Z}$. There will be times when we will need to refer to the elements of $\mathbb{Z}_n$, and since these are equivalence classes and not just integers, the notation must be chosen carefully. So we introduce the following: for $m \in \mathbb{Z}$, we write $[m]_n = \{x \in \mathbb{Z} : n|(x - m)\}$.

Note that we now have two ways of denoting exactly the same thing. For integers a, b, and n with $n > 1$, the two statements "$a \equiv b \pmod{n}$" and "$[a]_n = [b]_n$" are equivalent.

Let us stop and think about what this all means in the context of time. Suppose we are told that, "in this camp, breakfast is served at 7:00 A.M." What does this mean? Is there exactly one instant on a certain day and time at which breakfast is served? This can hardly be the case, since we eat breakfast every day. What it must mean is that breakfast is served at 7:00 A.M. today, tomorrow, yesterday, and in a week. So 7:00 A.M. actually represents many different times, as long as the difference between that time and 7:00 A.M. is a multiple of 24 hours. Mathematically, this idea is expressed by an equivalence class. The class of 7 modulo 24 is the set of all integers that differ from 7 by a multiple of 24. Thus, $[7]_{24} = [31]_{24} = [-41]_{24} = \ldots = \{\ldots, -41, -17, 7, 31, 55, \ldots\}$. The "numbers" in modular arithmetic are sets of numbers.

Before exploring some of the properties of the integers modulo n, we need to learn a bit more about the integers themselves. You are no doubt familiar with these properties of the integers, but you may not know the rigorous definitions or the exact statements of the theorems. The first statement is simply about division of one integer by another.

Theorem 25.3 (Division algorithm).
Let m and n be integers with $n > 0$. Then there exist unique integers q and r such that $m = nq + r$ and $0 \le r < n$.

In plain English, the division algorithm says that for two integers, m and n, we can write m as a multiple of n plus what's left over. Of course, that's just the statement that q is the quotient and r the remainder when we divide m by n. You might not understand why we need to prove this—after all, you have been using it for a long time. But did you ever stop to think about what it really means and why it is true? You will prove this theorem when you solve Problems 25.16 and 25.17.

Let's move on to another old friend from the past. Given two numbers, say 28 and 42, what is the gcd (or greatest common divisor) of the two numbers? Using previous knowledge, you probably figured out without too much trouble that the answer is 14. But now that you have much more mathematical experience, we are able to ask (and answer) the more complicated questions of "how do we define gcd precisely?" and, "is there an algorithm to find its value?"

Define the **greatest common divisor** d (which we'll soon see is unique) of two integers m and n, where m and n are not both zero, to be the positive integer d that satisfies

(i) $d|m$ and $d|n$, and
(ii) if s is a positive integer such that $s|m$ and $s|n$, then $s|d$.

We denote the greatest common divisor of m and n by $\gcd(m, n)$. We say m and n are **relatively prime** if $\gcd(m, n) = 1$.

Exercise 25.4.
(a) What does condition (i) of the definition of the greatest common divisor really say?

(b) What does condition (*ii*) really say?

(c) We mentioned that the gcd of two numbers is unique. How would you try to prove this? ○

Exercise 25.5.

Find gcd$(-16, 40)$, gcd$(0, 45)$, and gcd$(-30, -27)$. ○

The next theorem tells us that the gcd always exists and is, as we promised, unique.

Theorem 25.6.

Let m and n be integers, not both zero. Then their greatest common divisor exists, is unique, and there are integers k and l such that gcd$(m, n) = km + ln$.

This theorem actually tells us more than the existence and uniqueness of the greatest common divisor. It tells us that the gcd can be expressed as the sum of multiples of the two numbers m and n. This fact is certainly not obvious, and it will turn out to be very useful. A sum of the form $km + ln$ where k and l are integers is called a **linear combination** of m and n. Theorem 25.6 is usually proved by first showing that gcd$(m, n) = km + ln$. The proof looks at the set A of all the linear combinations of m and n that yield a positive integer. Since A will be a nonempty set of positive integers, the well-ordering principle tells us that this set has a smallest element. It turns out that this element will satisfy both (*i*) and (*ii*) in the definition of greatest common divisor. Once we have shown this, we will still need to present an argument that there is no other integer that is also the gcd of m and n.

Proof.

Let m and n be two integers, not both zero. Consider the set $A = \{xm + yn : x, y \in \mathbb{Z} \text{ and } xm + yn > 0\}$. First we'll show that $A \neq \emptyset$. We know that $m, n \in \mathbb{Z}$, and so we may set $x = m$ and $y = n$. Since $m \neq 0$ or $n \neq 0$, we conclude that $xm + yn = m^2 + n^2 > 0$, and hence $m^2 + n^2 \in A$. Thus $A \neq \emptyset$. By the well-ordering principle, every nonempty set of positive integers has a smallest element, and we call this element d. Since $d \in A$, there exist $x_0, y_0 \in \mathbb{Z}$ such that $d = x_0 m + y_0 n$. We will show that $d = \gcd(m, n)$, proving two parts of the theorem, namely that a greatest common divisor exists, and that this divisor can be written in the form $km + ln$ for some $k, l \in \mathbb{Z}$.

Since $d \in A$, we know that $d > 0$. By the division algorithm (Theorem 25.3), we can write $m = qd + r$ with $q, r \in \mathbb{Z}$ and $0 \le r < d$. Then $r = m - dq = m - (x_0 m + y_0 n)q = (1 - x_0 q)m + (-y_0 q)n$, where $1 - x_0 q$ and $-y_0 q$ are integers. Now if $r > 0$, then r would be an element of A. But $r < d$ and d is the smallest element of A. This means that $r \notin A$. Hence it must be the case that $r = 0$; in other words, $d|m$. Exactly the same argument shows that $d|n$. Thus $d|m$ and $d|n$, and (i) in the definition of gcd holds.

Now suppose that s is a positive integer such that $s|m$ and $s|n$. Since $d = x_0 m + y_0 n$, we conclude that $s|d$. (You are asked to write out the details of this last step in Problem 25.1.) Hence (ii) also holds for d. We now know that a greatest common divisor exists and has the right form. It remains to show that it is unique.

So suppose that d and s are both greatest common divisors of m and n. Then, since d is a gcd, property (ii) of the definition says that $s|d$. On the other hand, s is a gcd, so $d|s$. We conclude that $s|d$ and $d|s$. Since both s and d are positive integers it follows (see Problem 25.2) that $s = d$, showing the uniqueness of the greatest common divisor. ∎

Incidentally, while the greatest common divisor d is unique, the integers x_0 and y_0, as defined in the proof, are not. Here is a simple example: Consider the integers $m = 6$ and $n = 9$. Then

$$\gcd(6, 9) = 3 = 2 \cdot 6 + (-1) \cdot 9 = (-1) \cdot 6 + 1 \cdot 9.$$

Unfortunately, the proof of Theorem 25.6 was not constructive; that is, it's a nice enough proof, but it doesn't really tell us how to find $\gcd(m, n)$. However, there is an algorithm to do just that—one that appeared in Euclid's *Elements* 2,300 years ago. The algorithm is appropriately called the *Euclidean algorithm* and you will learn to apply it in Problem 25.18 to calculate the gcd of two integers.

We now return to modular arithmetic. To get you back into the proper state of mind, we suggest that you reread (in the beginning of this chapter) what it means for two integers to be equivalent modulo n, where $n > 1$. Then work the following exercise:

Exercise 25.7.
Show that for integers m and n with $n > 1$, there exists an integer r satisfying $0 \leq r < n$ such that $m \equiv r \pmod{n}$. ○

One good thing about the integers is that we can perform basic algebraic manipulations on them, like adding, subtracting, and multiplying. Can we do this on $\mathbb{Z}_n$ also? The answer is yes, but we must first carefully define how these operations work on the equivalence classes that make up the set $\mathbb{Z}_n$. That's what we will do right after we work an example to remind you what it means for two equivalence classes modulo n to be the same.

Example 25.8.
For integers r, s, and n with $n > 1$, prove that $[r]_n = [s]_n$ if and only if there exists $k \in \mathbb{Z}$ such that $r - s = kn$.

Proof.
By Problem 10.5, $[r]_n = [s]_n$ if and only if $r \sim s$. Thus $[r]_n = [s]_n$ if and only if $r \equiv s \pmod{n}$. Hence $[r]_n = [s]_n$ if and only if there exists an integer k such that $r - s = kn$. ∎

Be sure to keep this fact in mind as you read on in the text, and especially as you work your way through Example 25.9, in which we will show that multiplication on $\mathbb{Z}_n$, as introduced below, is well-defined.

Fix an integer $n > 1$. Now $\mathbb{Z}_n$ is closely related to $\mathbb{Z}$, so we will try to modify the operations of $\mathbb{Z}$ so that they apply to $\mathbb{Z}_n$. For $r, s \in \mathbb{Z}$, define

$$[r]_n + [s]_n = [r + s]_n, \ [r]_n - [s]_n = [r - s]_n, \ \text{and} \ [r]_n \cdot [s]_n = [rs]_n.$$

Before going on, convince yourself that

$$[12]_5 + [7]_5 = [4]_5, \ [12]_5 - [7]_5 = [0]_5, \ \text{and} \ [12]_5 \cdot [7]_5 = [4]_5.$$

These definitions amount to defining three functions from $\mathbb{Z}_n \times \mathbb{Z}_n$ to $\mathbb{Z}_n$, and we need to show that they are well-defined. We will do this here for multiplication, and leave addition and subtraction to you in Problem 25.8.

Example 25.9.
Define $f : \mathbb{Z}_n \times \mathbb{Z}_n \to \mathbb{Z}_n$, by $f([r]_n, [s]_n) = [rs]_n$. Then f is a well-defined function that yields a multiplication on $\mathbb{Z}_n$.

"*Understanding the problem.*" To show that a function is well-defined, we need to prove two things. (i) We must show that f maps $\mathbb{Z}_n \times \mathbb{Z}_n$ into $\mathbb{Z}_n$. In other words, we must show that for every element x in $\mathbb{Z}_n \times \mathbb{Z}_n$, there exists an element y in $\mathbb{Z}_n$ such that $f(x) = y$. It is important to recall that when an element of a set can be written in a special form, as is the case with x in $\mathbb{Z}_n \times \mathbb{Z}_n$, we should take advantage of it. So if x is in $\mathbb{Z}_n \times \mathbb{Z}_n$, then there exist integers r and s such that $x = ([r]_n, [s]_n)$.

(ii) We must also show that, for x in $\mathbb{Z}_n \times \mathbb{Z}_n$, if $f(x) = y$ and $f(x) = z$, then $y = z$. Again, we will expect to use the special form of x, y, and z. Now x can be written as $([r]_n, [s]_n)$ for some integers r and s. Our function f is supposed to look at the pair of equivalence classes, choose an integer from each (they could be r and s, but don't have to be), multiply these together, and produce the resulting equivalence class.

What could possibly go wrong? Let us look at an example. We know that $[7]_6 = [19]_6$, because $6|(19-7)$. By the definition of multiplication in $\mathbb{Z}_6$, we write $[7]_6 \cdot [4]_6 = [28]_6$ and $[19]_6 \cdot [4]_6 = [76]_6$. The left sides of both equations are the same, so the right sides had better be the same as well, or we have a fatal problem on our hands. Are they the same? Our proof will need to show that the result of the multiplication operation is independent of the particular integers we used to represent the equivalence classes.

"*Devising a plan.*" To prove part (i), we have to show that f is defined for every element of $\mathbb{Z}_n \times \mathbb{Z}_n$, and yields an element in $\mathbb{Z}_n$. For part (ii), we let $x \in \mathbb{Z}_n \times \mathbb{Z}_n$ and suppose that $f(x) = y$ and $f(x) = z$. We must show that $y = z$. Now, we can assume that there exist integers r, s, u, and v such that $x = ([r]_n, [s]_n) = ([u]_n, [v]_n)$ and $y = f([r]_n, [s]_n) = [rs]_n$, while $z = f([u]_n, [v]_n) = [uv]_n$. So we must show that $[rs]_n = [uv]_n$. By Example 25.8, we know that this means that we must show that there exists an integer m such that

$rs - uv = mn$. How can we show that such an integer m exists? By using what we know, namely that $([r]_n, [s]_n) = ([u]_n, [v]_n)$. Looks like we are now ready to carry out our plan.

Proof.
Let $x \in \mathbb{Z}_n \times \mathbb{Z}_n$. Then $x = ([r]_n, [s]_n)$ for some $r, s \in \mathbb{Z}$. Hence, $rs \in \mathbb{Z}$, and therefore $[rs]_n \in \mathbb{Z}_n$. By the definition of f, we have $f(x) = [rs]_n$. Thus, f maps $\mathbb{Z}_n \times \mathbb{Z}_n$ to $\mathbb{Z}_n$.

Again let $x \in \mathbb{Z}_n \times \mathbb{Z}_n$. Consider two arbitrary representations of x, say $x = ([r]_n, [s]_n)$ and $x = ([u]_n, [v]_n)$. Then $f(x) = [rs]_n$ and $f(x) = [uv]_n$. We need to show that $[rs]_n = [uv]_n$. Since $([r]_n, [s]_n) = ([u]_n, [v]_n)$, the definition of ordered pair implies that $[r]_n = [u]_n$ and $[s]_n = [v]_n$. Thus

$$u - r = kn, \quad \text{for some } k \in \mathbb{Z}, \text{ and} \qquad (25.1)$$
$$v - s = ln, \quad \text{for some } l \in \mathbb{Z}. \qquad (25.2)$$

To show that $[rs]_n = [uv]_n$, we calculate

$$\begin{aligned} uv - rs &= uv - rv + rv - rs \\ &= (u - r)v + r(v - s) \\ &= knv + rln \qquad \text{(using equations (25.1) and (25.2))} \\ &= (kv + rl)n. \end{aligned}$$

Now $kv + rl \in \mathbb{Z}$ and, by Example 25.8, $[rs]_n = [uv]_n$. Thus f is well-defined, as desired. ∎

Exercise 25.10.
Define "modular exponentiation" as follows: For an integer $n > 1$ and $a, b \in \mathbb{Z}$, define $[a]_n^{[b]_n} = [a^b]_n$. Either prove that this operation is well-defined, or give an example to show that modular exponentiation is not well-defined. ○

The operations defined here are commutative, associative, have additive and multiplicative identities, satisfy the distributive property, and every element has an additive inverse. (See Problem 25.11.) Thus, they satisfy everything you might reasonably expect of an operation except for one thing: not every nonzero element has a multiplicative inverse. The following immediate consequence of Theorem 25.6 tells us something about reciprocals in $\mathbb{Z}_n$.

Corollary 25.11.
Let n be a positive integer with n > 1. Then for every integer a with $\gcd(a, n) = 1$, *there exists an integer b such that* $ab \equiv 1 \pmod{n}$.

Before you read the proof of the corollary, write out what it means to say $ab \equiv 1 \pmod{n}$.

Proof.
Since $\gcd(a, n) = 1$, Theorem 25.6 tells us that there exist $b, c \in \mathbb{Z}$ such that $ba + cn = 1$. Then $ba - 1 = (-c)n$ and $-c \in \mathbb{Z}$. Thus, $ab \equiv 1 \pmod{n}$. ∎

For an integer a to satisfy the hypothesis of this corollary, a needs to be relatively prime to the modulus n. Is it possible that we have two integers, a and b with $a \equiv b \pmod{n}$, such that one of the integers, say a, satisfies the hypothesis and the other one, b, does not? The answer to this query is no, as we see from the following lemma:

Lemma 25.12.
Let a, c, and n be integers with n > 1 and such that $a \equiv c \pmod{n}$. *Then* $\gcd(a, n) = 1$ *if and only if* $\gcd(c, n) = 1$. *Further, if b and d are integers such that* $ab \equiv 1 \pmod{n}$ *and* $cd \equiv 1 \pmod{n}$, *then* $b \equiv d \pmod{n}$.

The proof of this lemma requires the multiplication defined on $\mathbb{Z}_n$ earlier in this chapter, and the (easily checked) algebraic properties of this multiplication. (See Problem 25.11.)

Proof.
For the first part of the proof, we prove the contrapositive; that is, we prove that if $\gcd(c, n) \neq 1$, then $\gcd(a, n) \neq 1$. So assume that $\gcd(c, n) = k > 1$. Since $a \equiv c \pmod{n}$, we conclude that $a - c = ln$ for some $l \in \mathbb{Z}$. Hence $a = c + ln$. But $k|c$ and $k|n$, so $k|a$. By *(ii)* in the definition of gcd, we conclude that $k | \gcd(a, n)$. Thus $\gcd(a, n) > 1$. The converse is obtained by interchanging the roles of a and c.

For the second part of the proof, we will use our assumptions that $[a]_n[b]_n = [ab]_n = [1]_n$, $[c]_n[d]_n = [cd]_n = [1]_n$, and $[a]_n = [c]_n$. Thus, we calculate

$$[b]_n = [bcd]_n = [bad]_n = [abd]_n = [d]_n,$$

and we conclude that $b \equiv d \pmod{n}$. ∎

Taken together, the corollary and the lemma tell us that if an integer a is relatively prime to n, then there exists an integer b such that the equivalence classes satisfy $[a]_n \cdot [b]_n = 1$. So, for a relatively prime to n, the equivalence class has something that should remind you of a reciprocal. This leads to the following definition: For a, b, and $n \in \mathbb{Z}$ with $n > 1$, we call b a **reciprocal modulo** n of a if $ab \equiv 1 \pmod{n}$. The notation is $b \equiv a^{-1} \pmod{n}$.

Exercise 25.13.
(a) Find the reciprocals modulo 7 of $3, 5$, and 6.
(b) Which elements of $\mathbb{Z}_6$ have reciprocals modulo 6 and which ones do not? ○

The use of modular arithmetic is widespread. Every time you are on the web, your browser is likely to make your transactions secure using an encryption that is based on modular arithmetic. (Work Project 27.11 on codes to see one such use.) We motivated the ideas in the chapter using time and calculations modulo 24. If you schedule tasks by days of the week you probably want to calculate with modulus 7; if you are interested in a monthly schedule, the modulus is 12. In fact, now that we've mentioned it, you can surely think of many other times when you have used modular arithmetic.

Solutions to Exercises

Solution to Exercise (25.1).
(a) We defined $5 \equiv x \pmod{12}$ by $12 \mid (x - 5)$. Some possible values for x are: $5, 17, 125, -7, -115$. Some values that do not work are: $0, 7, 1200, -5, -12$.

(b) The equivalence $-3 \equiv 39 \pmod{x}$ is defined by $x|42$ where x is an integer greater than 1. The set of all positive factors greater than 1 of 42 is the set $A = \{2, 3, 6, 7, 14, 21, 42\}$. Any five integers from A will work. The five non-solutions must be chosen from the integers greater than 1 that are not in A.

Solution to Exercise (25.4).

(a) Condition (i) says that the greatest common divisor divides both integers. In other words, it is a statement about being a common divisor.

(b) Condition (ii) says that every other positive integer that divides both m and n also divides the $\gcd(m, n)$, and therefore is a factor of it. In other words, it is a statement about being the greatest.

(c) We will need to prove the uniqueness of the greatest common divisor. To do so, we will prove that if there are integers d_1 and d_2, both satisfying the definition of greatest common divisor, then $d_1 = d_2$.

Solution to Exercise (25.5).

We list the answers here: $\gcd(-16, 40) = 8$, $\gcd(0, 45) = 45$, and $\gcd(-30, -27) = 3$.

Solution to Exercise (25.7).

The integers m and n are given and $n > 1$. By Theorem 25.3, there are $q, r \in \mathbb{Z}$ such that $m = nq + r$ and $0 \le r < n$. Hence $m - r = nq$ for some $q \in \mathbb{Z}$. Thus $m \equiv r \pmod{n}$ and $0 \le r < n$.

Solution to Exercise (25.10).

This operation is not well-defined. Let $n = 5$. Then $2 \equiv 7 \pmod 5$. Now $[3]_5^{[2]_5} = [4]_5$ and $[3]_5^{[7]_5} = [2]_5$, but $4 \not\equiv 2 \pmod 5$.

Solution to Exercise (25.13).

(a) It is easy to check that $3^{-1} \equiv 5 \pmod 7$, $5^{-1} \equiv 3 \pmod 7$ and $6^{-1} \equiv 6 \pmod 7$.

(b) By Corollary 25.11, the integers 1 and 5 have reciprocals modulo 6; it is easy to check that none of the others does.

Problems

Problem 25.1.
Let a, b, c, x, and $y \in \mathbb{Z}$. Prove that if $a|b$ and $a|c$ then $a|(bx + cy)$.

Problem 25.2.
Let a and b be positive integers such that $a|b$ and $b|a$. Prove that $a = b$.

Problem 25.3.
Prove Theorem 25.2. (Note that this generalizes part (c) of Problem 10.1.)

Problem 25.4.
Carefully read the definition of greatest common divisor. What should the least common multiple of two integers be? Make up a definition for it. The least common multiple of two integers m and n is denoted by $\mathrm{lcm}(m, n)$.

Problem 25.5.
Using your definition from Problem 25.4 and the notation defined above, prove that if m and n are positive integers, then

$$\gcd(m, n) \cdot \mathrm{lcm}(m, n) = mn.$$

Problem 25.6.
Let $m, n \in \mathbb{Z}$, not both zero. Suppose that a is an integer that divides both m and n and whenever s is an integer dividing both m and n, then $s \leq a$. Prove that $a = \gcd(m, n)$.

Problem 25.7.
Let $m, n \in \mathbb{Z}$ and assume that $m \neq 0$. Prove the following statements.
(a) For all positive integers k, we have $\gcd(mk, nk) = k \gcd(m, n)$.
(b) If $d = \gcd(m, n)$ and $k, l \in \mathbb{Z}$, then $\gcd(m, n)| \gcd(m+kd, n+ld)$.

Problem 25.8.
Let $n > 1$ be an integer.

(a) Define $g : \mathbb{Z}_n \times \mathbb{Z}_n \to \mathbb{Z}_n$ by $g([r]_n, [s]_n) = [r + s]_n$. Prove that g is well-defined.

(b) Define $h : \mathbb{Z}_n \times \mathbb{Z}_n \to \mathbb{Z}_n$ by $h([r]_n, [s]_n) = [r - s]_n$. Prove that h is well-defined.

Problem 25.9.
Define $f : \mathbb{Z}_{12} \to \mathbb{Z}_{24}$ by $f([x]_{12}) = [3x]_{24}$. Is f well-defined? Prove your claim.

Problem 25.10.
Let p be an odd prime and define $f : \mathbb{Z}_p \to \mathbb{Z}_p$ by $f([x]_p) = [2x]_p$. Prove that f is well-defined and that f is a bijection.

Problem♮ 25.11.
Let $n > 1$ be an integer. Using the addition and multiplication defined on $\mathbb{Z}_n$ in this chapter, prove the following statements:

(a) $([a]_n + [b]_n) + [c]_n = [a]_n + ([b]_n + [c]_n)$ for all $a, b, c \in \mathbb{Z}$;

(b) there is an integer $e \in \mathbb{Z}$ such that
   (i) $[a]_n + [e]_n = [e]_n + [a]_n = [a]_n$ for all $a \in \mathbb{Z}$, and
   (ii) for every $a \in \mathbb{Z}$, there exists $b \in \mathbb{Z}$ such that $[a]_n + [b]_n = [b]_n + [a]_n = [e]_n$;

(c) $[a]_n + [b]_n = [b]_n + [a]_n$ for all $a, b \in \mathbb{Z}$;

(d) $([a]_n \cdot [b]_n) \cdot [c]_n = [a]_n \cdot ([b]_n \cdot [c]_n)$ for all $a, b, c \in \mathbb{Z}$;

(e) $[a]_n \cdot ([b]_n + [c]_n) = [a]_n \cdot [b]_n + [a]_n \cdot [c]_n$ and $([a]_n + [b]_n) \cdot [c]_n = [a]_n \cdot [c]_n + [b]_n \cdot [c]_n$ for all $a, b, c \in \mathbb{Z}$;

(f) there is $u \in \mathbb{Z}$ such that $[a]_n \cdot [u]_n = [u]_n \cdot [a]_n = [a]_n$ for all $a \in \mathbb{Z}$;

(g) $[a]_n \cdot [b]_n = [b]_n \cdot [a]_n$ for all $a, b \in \mathbb{Z}$.

(A set with two well-defined operations satisfying (a)–(g) is called a "commutative ring with identity.")

Problem 25.12.
Let n be an integer, where $n > 1$. Prove that every element of $\mathbb{Z}_n$ has a reciprocal modulo n if and only if n is prime.

(This makes $\mathbb{Z}_p$, where p is prime, as good a set to do arithmetic in as $\mathbb{Q}$. As far as multiplication is concerned, $\mathbb{Z}_p$ is better than $\mathbb{Z}$, because very few numbers (two, to be exact) in $\mathbb{Z}$ have reciprocals that also lie in $\mathbb{Z}$. The set, $\mathbb{Z}_p$, for p a prime, falls under the heading of "fields," and so do $\mathbb{Q}, \mathbb{R}$, and $\mathbb{C}$, but $\mathbb{Z}$ does not.)

Problem 25.13.
(This problem is appropriate only if you studied Chapter 21.) Use the result of Exercise 25.7 to show that $|\mathbb{Z}_n| = n$.

Problem 25.14.
Find *all* solutions in $\mathbb{Z}_n$ for the following equivalences:
   (a) $3x \equiv 0 \pmod{12}$;
   (b) $3x \equiv 0 \pmod{17}$;
   (c) $3x \equiv 0 \pmod{10}$.

Problem 25.15.
Find *all* solutions in $\mathbb{Z}_n$ for the following equivalences:
   (a) $4x \equiv 1 \pmod{11}$;
   (b) $4x \equiv 1 \pmod{9}$;
   (c) $3x \equiv 1 \pmod{11}$;
   (d) $3x \equiv 1 \pmod{9}$.

Problem 25.16.
Let $m, n \in \mathbb{Z}$ with $n > 0$. Define the set $A = \{m - nx : x \in \mathbb{Z}$ and $m - nx \geq 0\}$.
   (a) What is A, if $m = 2$ and $n = 1$?
   (b) What is A, if $m = 18$ and $n = 5$? What is the smallest integer in A?
   (c) Prove that $A \neq \emptyset$.
   (d) Since A is a nonempty subset of the natural numbers, it contains a smallest integer. Call this integer r. Prove that $0 \leq r < n$.
   (e) Use your work in part (d) to prove that there are integers q and r such that $m = nq + r$ and $0 \leq r < n$.
(Note: This proves the existence part of Theorem 25.3.)

Problem 25.17.
Let $m, n \in \mathbb{Z}$ with $n > 0$. Suppose there are integers q_1, q_2, r_1, and r_2 such that $m = nq_1 + r_1$ and $m = nq_2 + r_2$, where $0 \leq r_1 < n$ and $0 \leq r_2 < n$. Prove that this implies that $q_1 = q_2$ and $r_1 = r_2$. (Note: This proves the uniqueness part of Theorem 25.3.)

Problem 25.18.

Here's a brief explanation of the Euclidean algorithm, which is an effective way to find the greatest common divisor of two integers m and n, not both zero. This algorithm is in the seventh book of Euclid's *Elements*, but was likely known earlier.

There are two trivial cases that must be considered before moving to the interesting one. If $m = n$, then the greatest common divisor is obviously $|m|$. If one of the integers is zero (remember that both can't be zero), then the greatest common divisor is the absolute value of the non-zero integer. Now for the main case, note that the positive divisors of an integer m are the same as the ones of $-m$. For this reason, we may assume that both m and n are positive. After possible relabelling of the two numbers, we may further assume that $m > n > 0$.

The Euclidean algorithm is a repeated application of the division algorithm, Theorem 25.3. Each line is obtained from the previous one by shifting the divisor to the spot previously occupied by the dividend, and the remainder to the spot previously occupied by the divisor. It's easier to see than to say. Here is the way to see it:

$$
\begin{aligned}
m &= q_1 n + r_1, \\
n &= q_2 r_1 + r_2, \\
r_1 &= q_3 r_2 + r_3,
\end{aligned}
$$

$$\ldots$$

$$
\begin{aligned}
r_{k-3} &= q_{k-1} r_{k-2} + r_{k-1}, \\
r_{k-2} &= q_k r_{k-1} + r_k, \\
r_{k-1} &= q_{k+1} r_k.
\end{aligned}
$$

By the division algorithm, the remainders satisfy the inequalities $n > r_1 > \ldots > r_i > r_{i+1} > \ldots > 0$. This guarantees that the algorithm comes to a halt after finitely many steps. We label the last non-zero remainder r_k and solve for r_k as follows:

$$
\begin{aligned}
r_k &= r_{k-2} - q_k r_{k-1} \\
&= r_{k-2} - q_k(r_{k-3} - q_{k-1} r_{k-2}) = -q_k r_{k-3} + (1 + q_k q_{k-1}) r_{k-2}
\end{aligned}
$$

$$\ldots$$

$$
= x_0 m + y_0 n
$$

It can be shown (but we won't ask you to do it) that $r_k = \gcd(m, n)$.

We'll work out one example for you, so you can see how this is done. We will find the greatest common divisor of 8 and 27 and express it as a linear combination of the given integers. Now we need $m > n$, so $m = 27$ and $n = 8$. We now proceed with the algorithm. The remainders are underlined, and will be replaced with what we obtained in the column on the left.

$$
\begin{array}{l|l}
27 = 3 \cdot 8 + \underline{3} & \text{so } \underline{1} = 3 - 1 \cdot \underline{2} \\
8 = 2 \cdot 3 + \underline{2} & \quad = 3 - 1 \cdot (8 - 2 \cdot 3) \quad = -8 + 3 \cdot \underline{3} \\
3 = 1 \cdot 2 + \underline{1} & \quad = -8 + 3(27 - 3 \cdot 8) = 3 \cdot 27 - 10 \cdot 8 \\
2 = 2 \cdot 1 &
\end{array}
$$

So our algorithm tells us that $1 = 3 \cdot 27 - 10 \cdot 8$, and you can now check that this answer is correct.

You'll understand the algorithm better if you use it to calculate the gcd of two numbers. Do so for the following pairs of integers (m, n) and find the corresponding integers x_0 and y_0:

(a) $(2745, 135)$;
(b) $(528, 627)$;
(c) $(4746, 894)$.

Problem 25.19.
Use the Euclidean algorithm of Problem 25.18 to show that 2542 and 4095 are relatively prime.

Problem 25.20.
On a calculator or a computer, program the Euclidean algorithm as outlined in Problem 25.18. Check your program by trying it out on parts (a) through (c) in that problem.

Problem 25.21.
In the text we defined what it means for an integer p to be prime. We also defined what it means for two integers a and b to be relatively prime. Give an alternate definition for an integer p to be prime by requiring a and p to be relatively prime for certain integers a. Prove that the original and the alternate definition of prime are equivalent.

Problem 25.22.

Let a, b, and p be integers and p a prime. Prove that if $p|ab$, then $p|a$ or $p|b$.

Problem 25.23.

It is possible to define a function f that tells you the day of the week your birthday will fall on each year. To construct such a function, you need to find out what day of the week you were born. (Encode the weekdays as: 0—Sunday, 1—Monday, and so on.) Letting s denote the encoded week day of your birth, a the year you were born, and b the year in which you want to know the week day of your birthday, you will need to define f in terms of s, a, and b. Thus, the required function f will be a map from $\mathbb{Z}_7 \times \mathbb{Z} \times \mathbb{Z}$ into $\mathbb{Z}_7$. (The formula will depend on whether your birthday is before, after, or on February 29 of your birth year.)

Use modular arithmetic but keep in mind that there are leap years. (The year 2000 was a leap year. The formula becomes considerably more complicated if you want to extend it past 2100, because that year will not be a leap year.)

To find out the day of your birth and to check your formula, access one of the perpetual calendars on the web such as [29].

26

Fermat's Little Theorem

CHAPTER

We begin this chapter with a fundamental result of number theory, discovered by Pierre de Fermat. Fermat lived from 1601 to 1665. Many of his contemporaries were "number-lovers" rather than number theorists, [83, p. 51], and one thing that interested them was perfect numbers (a number is perfect if it is the sum of all its proper divisors). Bernard Frenicle de Bessy, who was also a mathematician and physicist, first raised the question of whether there was a perfect number of 20 digits and, if not, what the next largest perfect number was. (See [24] and [25].) The answer to the question required determining whether certain large numbers were prime. Thus began a correspondence between the two men. In a letter to Frenicle, dated October 18, 1640, Fermat stated what is now known as Fermat's theorem or Fermat's little theorem (to distinguish it from Fermat's last theorem), but he did not include a proof. In 1736, almost a century later, Leonhard Euler gave the first rigorous proof of the little theorem. Though this theorem is clearly theoretical in nature, it plays an important role in primality testing; that is, in deciding whether or not a certain number is prime. Fermat's little theorem is also the mathematical heart of the widely used RSA code that we will describe later in this chapter. In fact, Fermat's little theorem is not little at all.

Theorem 26.1 (Fermat's Little Theorem).
Let p be a prime and let a be an integer satisfying $\gcd(a, p) = 1$. Then

$$a^{p-1} \equiv 1 \pmod{p}.$$

Exercise 26.2.
Verify Theorem 26.1 for a few values of p and a. ○

We will state and prove Euler's generalization of this theorem below. Fermat's little theorem will then follow as a special case.

In order to state Euler's generalization of Fermat's theorem, we have to introduce something that is now referred to as **Euler's ϕ-function**: For $n \in \mathbb{Z}^+$, let $\phi(n)$ be the number of integers k, with $0 \le k < n$, that are relatively prime to n.

Exercise 26.3.
Calculate $\phi(1)$, $\phi(12)$, $\phi(7)$, $\phi(13)$, and $\phi(7 \cdot 13)$. ○

Exercise 26.4.
Show that if p is prime, then $\phi(p) = p - 1$. ○

The following lemmas will assist us in our proof of Theorem 26.7 below. The first lemma requires the multiplication in $\mathbb{Z}_n$ that we defined in Chapter 25.

Lemma 26.5.
Let n and a be integers satisfying $n > 1$ and $\gcd(a, n) = 1$. If r and s are integers satisfying $ar \equiv as \pmod{n}$, then $r \equiv s \pmod{n}$.

Proof.
Since we know that $\gcd(a, n) = 1$, we may apply Corollary 25.11 to obtain an integer b such that $ab \equiv 1 \pmod{n}$. We now multiply the equivalence $ar \equiv as \pmod{n}$ by b to get $arb \equiv asb \pmod{n}$. Using commutativity of the multiplication (see Problem 25.11) and simplifying, we obtain $r \equiv s \pmod{n}$, as desired. ∎

We summarize much of what we have learned below.

Lemma 26.6.
Let a and n be integers with $n > 1$ and $\gcd(a, n) = 1$. Then there exist exactly $\phi(n)$ distinct integers, $m_1, m_2, \ldots, m_{\phi(n)}$ such that
(i) $0 \le m_i < n$ and $\gcd(m_i, n) = 1$ for $i = 1, \ldots, \phi(n)$,
(ii) there exists $c \in \mathbb{Z}$ such that

$$\left(\prod_{i=1}^{\phi(n)} m_i\right) c \equiv 1 \pmod{n}, \text{ and}$$

(iii) $am_i \not\equiv am_j \pmod{n}$ for $i \ne j$.

Proof.
By the definition of Euler's ϕ-function, there exist exactly $\phi(n)$ distinct integers satisfying (i).

Property (i) and Problem 26.1, imply that $\gcd(\prod_{i=1}^{\phi(n)} m_i, n) = 1$. Thus we may apply Corollary 25.11 to obtain an integer c such that

$$\left(\prod_{i=1}^{\phi(n)} m_i\right) c \equiv 1 \pmod{n},$$

completing the proof of (ii).

For part (iii) recall that $\gcd(a, n) = 1$. Now $m_1, \ldots, m_{\phi(n)}$ are distinct integers with $0 \le m_k < n$ for each k. So, if $i \ne j$, then $m_i \not\equiv m_j \pmod{n}$. Thus (the contrapositive of) Lemma 26.5 implies that $am_i \not\equiv am_j \pmod{n}$ for $i \ne j$. ■

Now we are ready for Euler's generalization of Fermat's little theorem.

Theorem 26.7 (Euler's Theorem).
Let a and n be integers with $n > 1$. If $\gcd(a, n) = 1$, then

$$a^{\phi(n)} \equiv 1 \pmod{n}.$$

Proof.
Let $m_1, m_2, \ldots, m_{\phi(n)}$ be as in Lemma 26.6. Then the $\phi(n)$ integers $am_1, am_2, \ldots, am_{\phi(n)}$ are distinct $\pmod{n}$, and $\gcd(m_i, n) = 1$.

Note that

$$a^{\phi(n)} \prod_{i=1}^{\phi(n)} m_i = \prod_{i=1}^{\phi(n)} (am_i). \tag{26.1}$$

Thus, if we can find the product $\prod_{i=1}^{\phi(n)} (am_i)$ as well as the reciprocal of $\prod_{i=1}^{\phi(n)} m_i$ (mod n), we can also compute $a^{\phi(n)}$ (mod n).

To compute the first product, use Exercise 25.7 to obtain $\phi(n)$ integers, $s_1, s_2, \ldots, s_{\phi(n)}$ such that $s_i \equiv am_i$ (mod n) and $0 \le s_i < n$. Now, $\gcd(a, n) = 1$ and $\gcd(m_i, n) = 1$, so by Problem 26.1, $\gcd(am_i, n) = 1$. Thus, Lemma 25.12 implies that $\gcd(s_i, n) = 1$. We therefore have found $\phi(n)$ different integers, $s_1, s_2, \ldots, s_{\phi(n)}$, all relatively prime to n. So $s_1, s_2, \ldots, s_{\phi(n)}$ is simply a (possible) reordering of $m_1, m_2, \ldots, m_{\phi(n)}$. Consequently

$$\prod_{i=1}^{\phi(n)} (am_i) \equiv \prod_{i=1}^{\phi(n)} s_i \equiv \prod_{i=1}^{\phi(n)} m_i \quad (\text{mod } n). \tag{26.2}$$

Combining (26.1) and (26.2) we get

$$a^{\phi(n)} \prod_{i=1}^{\phi(n)} m_i \equiv \prod_{i=1}^{\phi(n)} m_i \quad (\text{mod } n). \tag{26.3}$$

By Lemma 26.6, there is an integer c such that $(\prod_{i=1}^{\phi(n)} m_i)c \equiv 1$ (mod n). So, multiplying both sides of (26.3) by c, we obtain

$$\left(a^{\phi(n)} \prod_{i=1}^{\phi(n)} m_i \right) c \equiv \left(\prod_{i=1}^{\phi(n)} m_i \right) c \quad (\text{mod } n).$$

Using associativity and simplifying, we obtain $a^{\phi(n)} \equiv 1$ (mod n), as desired. ∎

Since Fermat's little theorem is a special case of Euler's theorem, with $n = p$ for a prime p and $\phi(p) = p - 1$, we now have a proof of Theorem 26.1 as well.

One interesting application of Euler's theorem is in an area of mathematics known as coding theory. Here's the idea: Suppose you want to transmit a message to a receiver, whom we shall refer to as Henry, in such a way that no one else can read it. This is done all the time. (Just think how often you have sent your credit card number over the internet!) The idea is to use a code that is difficult to decode. But of course, if it's too difficult, Henry won't be able to decode it either. Applying the code to our secret message is like applying a function. Henry needs to undo the code, or mathematically speaking, apply the inverse function. So we need something like a function that has an inverse, but whose inverse is very difficult to find. Such functions are called trapdoor functions. (Anyone can get in, but only Henry can get out.) Since it is virtually impossible to find the inverse function, the original function used to hide the message can be made public. A method that does that is called a public key encryption.

One particular trapdoor function leads to the following method. Henry, the receiver of the messages, decides on a function that is determined by the two integers, n and e, called the key of the code. Anybody who is interested can learn about these two numbers (this is why it is a *public key encryption*). If you want to send a message to Henry, then you first turn the English text into a positive integer m, called the plaintext. There are standard ways to do this, and it does not yet hide the message. (If the translation leads to a number m that is greater than n, the message must be divided into several smaller messages.) The plaintext, m, must now be scrambled so that its meaning cannot be deciphered by anyone except Henry. Or, mathematically speaking, we have to apply the trapdoor function to it. A simple but very safe way to do this, is to change m to m^e (mod n). It's interesting to note that though it appears that everyone has all the information Henry has, it turns out that Henry knows something no one else knows. We'll explain this once we tell you how Henry will unscramble the message. So the question is: How can Henry recover m from m^e (mod n)? It turns out that he will use Euler's theorem. Here's how:

Example 26.8.

Let m, n, and e be positive integers and suppose that $n > 1$, $\gcd(m, n) = 1$, and $\gcd(e, \phi(n)) = 1$. Find a positive integer d such that $(m^e)^d \equiv m$ (mod n).

Note that if Henry had d, he would have m^e, n, e, and d. According to this example he could then calculate $(m^e)^d$, which is equivalent to

m modulo n. In order for the solution to this problem to be useful, we need a constructive way to find d. We claim that (i) we can find an integer d such that $e \cdot d \equiv 1 \pmod{\phi(n)}$ and that (ii) any such integer will fulfill the requirement $(m^e)^d \equiv m \pmod{n}$.

Now since e is a positive integer relatively prime to $\phi(n)$, Theorem 25.6 guarantees the existence of integers k and l such that $1 = ke + l\phi(n)$. Let d be the smallest positive integer such that $d \equiv k \pmod{\phi(n)}$. Then $1 \equiv de \pmod{\phi(n)}$, which is what we needed to show.

For part (ii) of the claim, calculate $(m^e)^d = m^{ed}$, and recall that $m^{\phi(n)} \equiv 1 \pmod{n}$, by Euler's theorem. We just showed that $1 = ed + j\phi(n)$ for some $j \in \mathbb{Z}$, so

$$m^{ed} = m^{1-j\phi(n)} = m \cdot (m^{\phi(n)})^{-j} \equiv m \cdot 1 \equiv m \pmod{n}.$$

Note that in Problem 25.18 we gave a constructive method to find the integers k and l used above, and in Exercise 25.7, you showed the existence of the integer d. ◯

Now back to Henry. Remember that he has determined his n and e and has given out these two integers. He also calculated the very important integer d from Example 26.8, but kept it a secret. Now you may well be asking the question, "Why can't everyone with access to n and e calculate d themselves, and then read the messages meant for Henry?" The reason is that in order to find d, a person needs to know the modulus that determines d, namely $\phi(n)$. Henry knows (as you will once you work Problem 26.6) that if he chooses n carefully, such that it is the product of two primes p_1 and p_2, then $\phi(n) = (p_1 - 1)(p_2 - 1)$. So he lets p_1 and p_2 be two primes, each about 100 digits long. (He must be a little careful choosing p_1 and p_2, but we will not go into that here.) Now he and everyone else knows the product n, but not p_1 and p_2. This is the trapdoor. Henry knows n, p_1, and p_2. So he can find $\phi(n)$. But everyone else only knows n, so they would have to find p_1 and p_2. It takes no time at all to multiply two 100-digit numbers, but you cannot factor the product in a million years, not even with supercomputers! Henry's method to get secure messages is called the RSA public key encryption, and Example 26.8 is the mathematical content of it. To learn more about this ingenious and widely-used method, work Project 27.11 on Coding Theory.

Solutions to Exercises

Solution to Exercise (26.2).
We calculate two examples.
1. Let $p = 5$ and $a = 7$. Then $7^4 = 2401 \equiv 1 \pmod 5$.
2. Let $p = 13$ and $a = 8$. Then $8^{12} = 68719476736$.
   Now $8^{12} - 1 = 68719476735 = 13 \cdot 5286113595$. Hence $8^{12} \equiv 1$
   (mod 13). ○

Solution to Exercise (26.3).
$\phi(1) = 1$.
  The nonnegative integers smaller than 12 are all listed. We cross out the ones that are not relatively prime to 12: $\emptyset$, 1, $\cancel{2}$, $\cancel{3}$, $\cancel{4}$, 5, $\cancel{6}$, 7, $\cancel{8}$, $\cancel{9}$, $\cancel{10}$, 11. Hence $\phi(12) = 4$.
  Similarly, $\phi(7) = 6, \phi(13) = 12$, and $\phi(7 \cdot 13) = \phi(91) = 72$.
  Notice that, in these examples, for p prime $\phi(p) = p - 1$, and $\phi(7 \cdot 13) = \phi(7) \cdot \phi(13)$. Are these coincidences?

Solution to Exercise (26.4).
Note that for p prime and a an integer with $0 \leq a < p$, we have $\gcd(a, p) = 1$ if and only if $a \neq 0$. Thus $\phi(p) = p - 1$ for every prime p.

Spotlight: Public and Secret Research

Research in mathematics today is often done by professors who work at universities or colleges. People frequently work collaboratively, though they also sometimes work alone. They might communicate via e-mail, get together when they can, work together at institutes, or they may never even meet each other. Once their work is done, they write it up and send it to a journal. The editor of the journal sends it to carefully selected referees who read the paper. The author is responsible for the correctness of the mathematics in the paper, but the referee (whose identity is generally hidden from the author) determines the value of the work, the appropriateness of its placement in the journal, the originality of the mathematics, and often the correctness of the results. Once the paper appears, everyone has access to the results and proofs in the paper.

There are also other places where mathematical research is done. In the United States, the *National Security Agency (NSA)* refers to itself as the "leading employer" of non-academic mathematicians. In Great Britain, there is the *Government Communications Headquarters (GCHQ)*, the successor to the famous Bletchley Park where British code breakers were so successful in intercepting and reading Nazi attack plans. The mathematics done in a place like this might become the government's secret, and therefore may never be published. Public key encryption is an example of how such secrecy may hamper mathematical progress.

In 1976, Whitfield Diffie, Martin Hellman, and Ralph Merkle developed the idea of the public key. In 1977, Ronald Rivest, Adi Shamir, and Leonard Adleman gave us the RSA code. A couple of decades later, it became known that British mathematicians at GCHQ had worked out the encryption idea a few years earlier. James Ellis, Clifford Cocks, and Malcolm Williamson, all employed at GCHQ, had discovered public key encryption, but neither they nor their supervisors realized the power and widespread applicability of the method. Their results were considered top secret and were only circulated within the agency. A few years later, the world admired the "new" cryptosystem and celebrated the "originators" in the United States, [78, Chapter 6].

Some private companies also restrict their employees' publications. There are instances of researchers circumventing this restriction by publishing under a penname. In 1908, William S. Gosset who worked for the Guiness Brewing Company in Dublin, published a paper under the name "Student" to avoid repercussions. The distribution Gosset introduced is now known as *Student's t-distribution* (or the Student t-distribution) and it has had a profound impact on statistical theory and practice.

To learn more about the interesting history of public key encryption we recommend [78, Chapter 6] or [50].

An in-depth treatment of Fermat's little theorem, Euler's theorem and Euler's ϕ-function, as well as historical notes can be found in the text [12, pp. 91-96 and 123-150]. In [46, pp. 418-420 and 556-558]

Fermat's theorem and Euler's theorem are put in context. For short biographies of Pierre de Fermat and Leonhard Euler, see the web at [74]. For a biographical sketch of Euler and a delightful description of Euler's mathematics in the various fields, see [19]. A good source to learn more about Fermat's life and his mathematics is [54].

Problems

Problem$^\sharp$ 26.1.
(a) Let a, b, and s be integers such that $\gcd(a, s) = 1$ and $\gcd(b, s) = 1$. Show that $\gcd(ab, s) = 1$.
(b) Let $n \in \mathbb{Z}^+$ and $a_1, \ldots, a_n$, and s be integers such that $\gcd(a_k, s) = 1$ for all integers k with $1 \le k \le n$. Show that $\gcd(\prod_{k=1}^n a_k, s) = 1$.

Problem 26.2.
Show that the conclusion of Fermat's little theorem (Theorem 26.1) may not hold if p is not prime.

Problem 26.3.
(a) Calculate $\phi(5^2)$, $\phi(5^3)$, and $\phi(5^4)$.
(b) For p a prime and n a positive integer, show that $\phi(p^n) = p^n(1 - 1/p)$.
(c) Calculate $\phi(128)$.

Problem 26.4.
Is Euler's ϕ-function additive? In other words, is $\phi(m+n) = \phi(m) + \phi(n)$ for all $m, n \in \mathbb{Z}^+$? Prove it or give a counterexample.

Problem 26.5.
This problem guides you through the proof of the fact that Euler's ϕ-function is in some sense multiplicative. More precisely, you will prove

Theorem 26.9.
Let m and n be integers such that $m > 1$ and $n > 1$. If $\gcd(m, n) = 1$, then $\phi(mn) = \phi(m)\phi(n)$.

(a) Let k_1, k_2, l_1, and l_2 be integers satisfying $0 \leq k_1, k_2 < n$ and $0 \leq l_1, l_2 < m$. Show that if $k_1 m + l_1 n \equiv k_2 m + l_2 n \pmod{mn}$ then $k_1 = k_2$ and $l_1 = l_2$.

(b) Use the result of (a) to show that for each $a \in \mathbb{Z}$ there is exactly one element $(k, l) \in \mathbb{Z} \times \mathbb{Z}$ such that $0 \leq k < n$, $0 \leq l < m$, and $a \equiv km + ln \pmod{mn}$.

(c) From (b), conclude that $\phi(mn)$ is equal to the number of elements $(k, l) \in \mathbb{Z} \times \mathbb{Z}$ such that $0 \leq k < n$, $0 \leq l < m$, and $\gcd(km + ln, mn) = 1$.

(d) For integers k and l satisfying $0 \leq k < n$ and $0 \leq l < m$, show that $\gcd(km + ln, mn) = 1$ if and only if $\gcd(k, n) = 1$ and $\gcd(l, m) = 1$.

(e) Use (c) and (d) to obtain the conclusion of Theorem 26.9.

Problem 26.6.
Use the results of Problems 26.3 and 26.5 to answer the following.

(a) Let $m \in \mathbb{Z}$ and suppose $m = p_1^{a_1} \cdot p_2^{a_2} \cdot \ldots \cdot p_k^{a_k}$, where $p_1, p_2, \ldots, p_k$ are distinct primes and $a_1, a_2, \ldots, a_k$ are positive integers. Prove that

$$\phi(m) = m \prod_{i=1}^{k} \left(1 - \frac{1}{p_i}\right).$$

(b) Calculate $\phi(5712200)$.

Problem 26.7.
Prove that for $n \in \mathbb{Z}^+$, the Euler ϕ-function satisfies

$$\phi(2n) = \begin{cases} \phi(n) & \text{if } n \text{ is odd} \\ 2\phi(n) & \text{if } n \text{ is even} \end{cases}.$$

Problem 26.8.
Let a, b, and c be integers such that $\gcd(a, b) = 1$ and $a|bc$. Prove that this implies that $a|c$.

Problem 26.9.
Prove Theorem 26.1 directly by adapting the proof of Theorem 26.7 to this simpler situation.

Problem 26.10.
Use the method of Example 26.8 to find an integer x satisfying the equivalence
  (a) $x^3 \equiv 9 \pmod{33}$;
  (b) $x^{77} \equiv 15 \pmod{143}$.

Problem 26.11.
Show that the conclusion of Euler's Theorem may not hold if $\gcd(a, n) > 1$.

Problem 26.12.
Show that if $x, y \in \mathbb{Z}$ and p is a prime, then $(x + y)^p \equiv x^p + y^p$ $\pmod{p}$. (You may find the binomial theorem useful here. If so, you may use, without proof, the fact that the binomial coefficient $\binom{n}{k}$ is an integer for $k, n \in \mathbb{N}$ with $k \leq n$.)

27

Projects

Tips on Talking about Mathematics

It's not easy to talk about mathematics to other people. In this section, we present some tips that we find helpful when we present a talk to undergraduates.

Let's say someone has just asked you to give a talk about mathematics to undergraduates. Here's what you need to do:

- Thank them, and say you'd love to. Then do the rest of the things below.
- Find out who the audience is and what they know.
- Pick an interesting topic. Find out about the history of the topic, the main players in the field, and the main results.
- Now that you have your topic, you need to write the talk. Start with something everyone is interested in. This could be the history of what you plan to talk about, or it could be an interesting related result. Then motivate the question you are interested in looking at, build up the talk, and remember to find a good conclusion for it.
- As you write your talk, keep the level of the audience in mind. Do not use terms that your audience will not understand. If they

haven't heard certain words you will have to define them, which brings us to our next point: the more terms you have to define, the more people you will lose. Pick a topic that doesn't require a lot of introduction.

- You need to decide whether you will use transparencies, the computer, or the blackboard. Each has its advantages and disadvantages. We'll run through each below.

1. *Blackboard.* If you use the blackboard, you'll most likely move at the right speed for the audience. It's also livelier than the other methods. On the other hand, you should absolutely not rely on your notes. Therefore, if you give a talk using the blackboard, you'll need to know what you are going to say and when you are going to say it. You'll need to watch where you write things, and you shouldn't erase something you want the audience to look at. Make sure that you move away from the board so that everyone can see what you wrote. If your handwriting is illegible, think about using transparencies or the computer.

2. *Transparencies.* Unless you are very careful, you will probably move too quickly for the audience. You'll probably also stand in front of the transparency from time to time, blocking the audience's view. If you are aware of these potential problems, you can correct them. For example, you can use two overheads. You should not write too much on one transparency, and you should always be aware of where you are standing. Find out how big the room is, and make sure that someone in the back of the room will be able to see what you have written. The advantage of transparencies or the computer is that you'll have all your diagrams and pictures in place, and you'll have an outline of your talk with you. So the main disadvantages are that your talk may become monotonous and that it's possible to move so quickly that your audience won't be listening. These are pretty big disadvantages.

3. *Computer.* In many ways, using the computer to give your talk is similar to using transparencies. Many of the advantages and disadvantages are the same. You can liven up the talk by adding relevant photographs of places, manuscripts, or people.

You might even add a video clip. Just make sure that these "attention getters" are relevant and well incorporated.

4. *Blackboard, Transparencies, and Computer.* One thing you can do is combine two or three of these methods of presentation. In a talk for undergraduates, it's nice for them to have something to look at from time to time, other than the speaker.

Pick the method you are most comfortable with and that you like the best. Then work around the disadvantages.

- So now you have your topic, your talk, and a method of presentation. You're done, right? Um ... no. You still have to present the talk. Surprisingly, the hardest part of the talk is timing. We've alluded to this already in our discussion on transparencies, but there's more to be said.
- Find out how long the talk is. If it's twenty minutes, talk for twenty minutes. (No one will complain if it's eighteen minutes, and everyone will complain if it's thirty.) There is only one way to know how long your talk is: practice it.
- The best way to practice a talk is to give it to yourself once. Fix the things you realize need fixing. Then try to find an audience of two people, one who knows what you are talking about and one who does not. Ask them if you can present the talk to them. Listen to their comments and use them to improve your talk.
- Write an interesting, but truthful abstract. The abstract should indicate the level of the talk.
- Before you give your talk, ask if you can see the room that you will speak in. Check that everything you need is there.
- Make sure that everyone in the room can hear you when you speak. When you give the talk, look at the audience. They'll let you know how you are doing.

There are other articles on how to talk about mathematics ([55], [33]), but these are primarily aimed at graduate students or professional mathematicians. Of course, many of the tips are the same, because many of the mistakes people make—whether talking to undergraduates, graduate students, or professors—are the same.

27.1 Picture Proofs

Introduction

You have probably heard the saying "a picture is worth a thousand words." The same is true in mathematics: a good picture can help a reader visualize what is happening, it can aid a mathematician in finding a solution, and it can shed light on other potential results. A bad picture, on the other hand, can be deceiving. Relying too much on what we see might lead us to incorrect proofs, which in turn can lead to false results. This project should help convince you of that.

One of the most influential theorems in mathematics is Pythagoras' theorem. It states that in a right triangle the lengths of the sides of the triangle satisfy $a^2 + b^2 = c^2$, where c denotes the length of the hypotenuse, and a and b denote the lengths of the other two sides of the triangle. There are many known proofs of this theorem, some of them based on clever figures. You will see two such proofs below.

Prerequisites

Basic geometry skills plus an understanding of what constitutes a rigorous argument are the necessary prerequisites for this project. We suggest that you read through Chapter 5 before attempting this project.

Guided Project

1. The diagram of Figure 27.1 suggests a proof of Pythagoras' theorem. To make this proof rigorous, however, you will need to do two things; you need to prove something about the diagram and you need to do an algebraic calculation. Do both.
2. Give a second proof of Pythagoras' theorem based on Figure 27.2. This one does not need algebraic calculations. It is all in the picture—or is it?
3. If you accepted the picture of Figure 27.2 as a complete proof of Pythagoras' theorem, then you are probably willing to believe

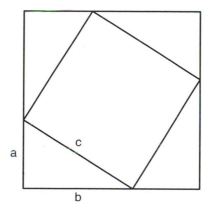

FIGURE 27.1 $a^2 + b^2 = c^2$

that Figure 27.3 provides a proof that $168 = 169$. After all, both proofs require that we shift the pieces around to form another familiar object. What is wrong with this proof?

4. Find a picture to illustrate the statement

$$\sum_{k=1}^{\infty} \frac{1}{2^k} = 1.$$

Does your picture amount to a rigorous proof? What are its strengths and what are its weaknesses?

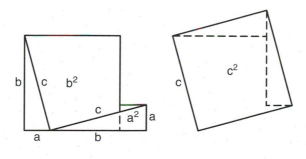

FIGURE 27.2 $a^2 + b^2 = c^2$

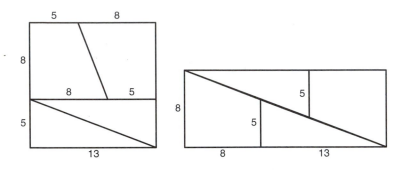

FIGURE 27.3 $169 = 13^2 = (8 + 13) \cdot 8 = 168$

Open-Ended Project

Try to find some other picture proofs. (We suggest you think back to your geometry course.) Can you also come up with a (somewhat) convincing picture proof of a false statement?

Notes and Sources

We first learned of the false proof presented in part 3 from our colleague, G. Adams. For a connection between this problem and Fibonacci sequences see the article [43], where the author indicates that this "not a picture proof" can be traced back to the year 1868. There are two excellent books on picture proofs by R. Nelsen, [59] and [60]. The website by A. Bogomolny [10] contains 39 proofs of Pythagoras' theorem, many of them with pictures, and some of them with applets.

27.2 The Best Number of All

Introduction

We know that positive integers can be even or odd, they can be prime or composite, and they can be triangular or square. (Each of these terms, with the exception of "square," appears either below or in the index of this book, if you have forgotten the definition.) In this project we'll look at some more things that they may be: the sum of their proper divisors, the product of their proper divisors, or both.

Prerequisites

This exercise requires an understanding of proof techniques (Chapter 5).

Guided Project

Recall that an integer greater than 1 is **prime** if its only positive divisors are 1 and itself. A positive integer greater than 1 that is not prime is called **composite**. By a proper divisor, we mean a positive divisor that is not equal to the integer itself. An integer is said to be **perfect** if it is the sum of its proper divisors.

1. Show that 6 is a perfect number.
2. Show that 6 is the only perfect number less than 10.
3. Find another perfect number that is less than 30.

By now you should have found the first two perfect numbers. The next is 496.

4. Check that 496 is a perfect number.
5. Find five positive integers, each one being the product of all of its proper divisors.
6. Characterize all positive integers that are the product of their proper divisors.

Now you are almost ready to prove the main theorem in this project:

Theorem 27.1.
There is only one positive integer that is both the product and sum of all its proper positive divisors, and that number is 6.

7. Let p be a prime. Prove that p^3 is not perfect.

8. Prove as many of the following as you need to, until you see the proof of the theorem.

   (a) Prove that the only even number that is both the product and sum of all its proper positive divisors is 6.
   (b) Prove that the only multiple of 3 that is both the product and sum of all its proper positive divisors is 6.
   (c) Prove that there is no multiple of 5 with this property.
   (d) Prove that there is no multiple of 7 with this property.

9. Prove the theorem.

Open-Ended Project

Let p be a polynomial over $\mathbb{Z}$ or $\mathbb{R}$ (see Problem 10.8). What might it mean to say that a polynomial is prime? composite? perfect? square? triangular? While some of these might make sense, others may not. Once you have defined the terms that make sense, are there some interesting theorems you can prove about them?

Notes and Sources

Euclid, in his *Elements*, IX.36, gave the result that if $p = 2^k - 1$ is a prime, then $2^{k-1}p$ is perfect. For $k = 2, 3, 5,$ and 7 we note that $2^k - 1$ is prime. Thus we get four perfect numbers, 6, 28, 496, and 8128. The eighth perfect number is already quite large: 2,305,843,008,139,952,128. A good source to begin learning more about numbers is the book [17]. The problem we discussed in the guided portion of this project appears in [12], which we recommend to those wanting to know more about number theory.

27.3   Set Constructions

Introduction

It is amazing how much one is able to build out of almost nothing— and by almost nothing here we mean the empty set. The guided project will lead you through a construction of the natural numbers.

Prerequisites

While the set theory introduced in Chapters 6–9 is sufficient, you may find it helpful to have an understanding of mathematical induction (Chapter 17), which is also introduced in this project.

Guided Project

Let x be a set. Define the **successor** of x to be the set $x^+ = x \cup \{x\}$.

1. Determine the successors, and the successors of the successors of the sets $\emptyset$, $\{\emptyset\}$, and $\{a, b, c\}$.

We now introduce the following notation. Let $0 = \emptyset$, $1 = 0^+$, $2 = 1^+$, and so on.

2. Write down $0, 1, 2, 3$, and 4 as sets in two different ways; first using the definitions made above, and then using only the symbol $\emptyset$, set brackets, and appropriate set notation.

It may seem intuitively obvious that if we "do this forever," then we will have defined the natural numbers. However, as simple and as attractive as this approach may be, it is not what we call mathematically rigorous. What we need is a statement that explains that we can do this forever. We will take this statement as an axiom, and thus it will not be proved.

Axiom 27.2 (Axiom of infinity).
There exists a set containing $\emptyset$ and the successor of each of its elements.

3. Let I be a nonempty set and $\{A_k : k \in I\}$ be an indexed family of sets. Suppose that for each $k \in I$, the set A_k has the two properties (i) $\emptyset \in A_k$ and (ii) if $x \in A_k$, then $x^+ \in A_k$. Show that the set $\bigcap_{k \in I} A_k$ also has these two properties. We will call a set with these two properties a successor set.
4. The axiom of infinity guarantees the existence of a successor set. So, let A be an arbitrary successor set. Define the set ω_A to be the intersection of all subsets of A that are also successor sets. In

symbols we might write

$$\omega_A = \bigcap_{B \in I} B,$$

where $I = \{B : B \subseteq A, \text{ and } B \text{ is a successor set}\}$.

By part 3, ω_A is a successor set. Show that $\omega_A = \omega_B$ for all successor sets A and B. Perhaps surprisingly, our definition does not depend on our initial choice of successor set, and therefore we will write ω rather than ω_A.

We call ω the set of natural numbers. Thus far we know that ω is a successor set, and it is the only successor set that is contained in every other successor set.

5. Prove the following statement. Suppose $S \subseteq \omega$ satisfies the two properties

   (i) $\emptyset \in S$, and
   (ii) if $x \in S$, then $x^+ \in S$.

   Show that $S = \omega$. (This is called the principle of mathematical induction and is discussed in Chapter 17.)
6. Prove that $x^+ \neq 0$ for all $x \in \omega$.
7. Consider the set $S = \{x \in \omega : \forall y \in \omega, \text{ if } y \in x, \text{ then } y \subseteq x\}$. Use part 5 to show that $S = \omega$. Conclude that for all u and v in ω, if $u \in v$, then $u \subseteq v$.
8. Use part 7 to prove that if x and y are in ω and $x^+ = y^+$, then $x = y$.

The two defining properties ((i) and (ii)) of a successor set, the principle of mathematical induction, and parts 6 and 8 of this project are known as the five Peano axioms. They are the pillars of the construction of the natural numbers.

Open-Ended Project

We have created new sets from old ones using element relations, union, intersection, power sets, and Cartesian products. Use some (or all) of these to create new sets from the empty set. Do your new sets have some interesting properties?

Notes and Sources

This project is guided by Chapters 11 and 12 of P. Halmos' *Naive Set Theory* [31]. For a brief presentation of the Peano axioms and some other attempts to give the natural numbers a solid foundation see [48, pp. 987–989].

27.4 Rational and Irrational Numbers

Introduction

We know that when we add two rational numbers, the result is a rational number. For this reason, we say that the rationals are **closed under addition**. Similarly, when we multiply two rational numbers, the result is rational. Thus, the rationals are also **closed under multiplication**. In this project, you will investigate the behavior of the rationals and irrationals under other operations.

Prerequisites

This project relies on proofs in cases (Chapter 5), as well as familiarity with rational and irrational numbers. In particular, you will need to use the fact that $\sqrt{2}$ is irrational.

Guided Project

Let a and b be two irrational numbers.

1. Give an example of two irrational numbers a and b such that $a+b$ is irrational.
2. Give an example of two irrational numbers a and b such that $a+b$ is rational.

So the irrational numbers are not closed under addition and certainly are less well behaved than the rational numbers. Now consider two real numbers, a and b.

3. Give an example of two rational numbers a and b such that a^b is rational.

4. Give an example of two rational numbers a and b such that a^b is irrational.

Here's a charming little proof, based entirely upon things that you have proved in this course, that an irrational number raised to an irrational power can be rational.

5. Consider the following.

Theorem 27.3.
There exist irrational numbers a and b such that a^b is rational.

Complete the proof of this theorem, using appropriate choices for a and b and the two cases below:

Case 1. $\sqrt{2}^{\sqrt{2}}$ is a rational number;

Case 2. $\sqrt{2}^{\sqrt{2}}$ is an irrational number.

The interesting thing about your proof of Theorem 27.3 is that you don't need to know whether $\sqrt{2}^{\sqrt{2}}$ is rational or irrational!

6. There are many other examples of irrational powers of irrational numbers that are rationals, assuming you know lots of different ways to express irrational numbers. See if you can come up with another example based on the fact that the natural logarithm of 2, denoted ln 2, is irrational. Can you find other examples?

Knowing that an irrational number to an irrational power may be rational raises the question of whether an irrational to an irrational can be irrational. Again, we are looking for a proof that does not use anything more than what we stated in the prerequisites. There are some nonelementary proofs of this, but an elementary proof exists as well [45].

7. Prove the following theorem using our suggestion (and the fact that $\sqrt{2}$ is irrational).

Theorem 27.4.
There exist irrational numbers a and b such that a^b is irrational.

The proof of this theorem is also a proof in cases. We suggest that you figure out how to use $\sqrt{2}^{\sqrt{2}}$ and the number $\sqrt{2}^{(\sqrt{2}+1)}$.

Open-Ended Project

Study the behavior of the rationals and irrationals under different operations. Your investigations might deal with specific numbers, or with the rationals and irrationals in general. For example, is $\sqrt{2}+\sqrt{3}$ irrational? In another direction, can you define an operation, $\odot$, such that $a \odot b$ is irrational for all irrational a and b? Think of other questions along these lines and try to answer them.

Notes and Sources

The connection of this problem to Hilbert's seventh problem is discussed in the Spotlight: Hilbert's Seventh Problem at the end of this chapter. The proof that an irrational number to an irrational power can be irrational appears in [45]. These authors attribute the proof of Theorem 27.3 to D. Jarden, [44]. This problem appears as a "fun fact" on the web at [37].

27.5 Irrationality of e and π

Introduction

The problems in this project require knowledge of calculus. More specifically, you need to know what the number e is, what a geometric series is, and what the series expansion for e is. If you have seen all this, then you probably have also been told that e is an irrational number. The first task of this project is to work through Ivan Niven's proof of this fact. If you have never seen the proof, it's a nice application of series. Everything you need to prove that e is irrational is provided in this project.

The proof that π is irrational, outlined in this project, is also due to I. Niven. In his words, "In the June 1947 issue of the *Bulletin of the A. M. S.*, I gave a one page proof that π is irrational. I had worked on this problem for a specific reason: in the first edition (1938) of what is now a great classic, *Introduction to the Theory of Numbers*, by G. H. Hardy and E. M. Wright, the authors made the observation that 'There is no simple proof of the irrationality of π.' I wondered why this should be so." (See [2] for the full text of Niven's conversation.)

We have provided you with all the steps you need to recreate Niven's one-page proof.

Prerequisites

Since the proofs are by contradiction, you will need to have covered Chapter 5. This project also assumes that you have a basic understanding of infinite series.

For the proof that these numbers are irrational, you will need to recall three results from your calculus course. The first is that, for $-1 < r < 1$, the geometric series satisfies

$$\sum_{n=0}^{\infty} r^n = \frac{1}{1-r}.$$

The second fact is that the series expansion for e^x is

$$e^x = 1 + x/1! + x^2/2! + x^3/3! + \cdots + x^k/k! + \cdots.$$

The last result that you will need is the product rule for differentiation.

Guided Project

1. Prove the following theorem, using the steps outlined below.

Theorem 27.5.
The number e is irrational.

Step 1. Let $k \in \mathbb{Z}^+$. Show that

$$\frac{1}{(k+1)} + \frac{1}{(k+1)(k+2)} + \frac{1}{(k+1)(k+2)(k+3)} + \cdots$$
$$\leq \frac{1}{(k+1)} + \frac{1}{(k+1)^2} + \frac{1}{(k+1)^3} + \cdots.$$

Step 2. Prove that if k is an integer with $k \geq 2$, then

$$\frac{1}{(k+1)} + \frac{1}{(k+1)(k+2)} + \frac{1}{(k+1)(k+2)(k+3)} + \cdots < 1.$$

Step 3. Suppose to the contrary that e is rational. Prove that this implies there exists an integer k such that $k!e$ is an integer.
Step 4. Using the series expansion for e, show that $k!e$ is never an integer. This step should complete the contradiction.
2. Prove the following theorem by contradiction, using the steps below.

Theorem 27.6.
The number π is irrational.

Suppose to the contrary that there are positive integers a and b such that $\pi = a/b$. We will write $f^{(m)}$ for the mth derivative of f. For $n \in \mathbb{Z}^+$, define the two polynomials f_n and F_n by

$$f_n(x) = \frac{x^n(a - bx)^n}{n!}, \quad \text{and}$$

$$F_n(x) = f_n(x) - f_n^{(2)}(x) + f_n^{(4)}(x) - \cdots + (-1)^n f_n^{(2n)}(x).$$

We will determine a value for n in the fifth step below. Until then, assume that n is a positive integer.
Step 1. Show that for every j, each of the following are integers: $f_n(0)$, $f_n(\pi) = f_n(a/b)$, $f_n^{(j)}(0)$ and $f_n^{(j)}(\pi) = f_n^{(j)}(a/b)$.
Step 2. Prove that $f_n(x) \sin x = \frac{d}{dx}(F_n'(x) \sin x - F_n(x) \cos x)$.
Step 3. Prove that $\int_0^\pi f_n(x) \sin x \, dx = F_n(\pi) + F_n(0)$.
Step 4. Find the maximum of the function f_n on the interval $[0, \pi]$. (Note that the maximum depends on n.)

Step 5. Prove that for n sufficiently large, $\int_0^\pi f_n(x) \sin x \, dx$ is not an integer. This step should complete the contradiction.

Open-Ended Project

Can you prove that e^2 is irrational? What else can you prove is irrational?

Notes and Sources

In 1737, Euler showed that e is irrational. Johann Heinrich Lambert showed, in 1761, that π is irrational. The number π has a very interesting history. For a brief history of π, see [22, p. 100]. For a fuller account, up to about 1971, see [8]. For more recent developments see [5].

The one-page proof in the *Bulletin of the A.M.S.* that Niven refers to in the quote above can be found in [61]. The reference for Hardy and Wright's text is given in [36]. The conversation with Niven appears in [2].

27.6 When Does $f^{-1} = 1/f$?

Introduction

Students often confuse the inverse of f, denoted f^{-1}, with the multiplicative inverse of f, denoted $1/f$. When are these two equal? Surprisingly, although the mistake of assuming $f^{-1} = 1/f$ is common, functions that have this seemingly intuitive property are not common at all.

Prerequisites

This project requires an understanding of functions and their inverses, presented in Chapters 13–15.

Guided Project

In what follows, f will always denote a real-valued function defined on a subset of $\mathbb{R}$ and satisfying $f^{-1} = 1/f$.

1. What can you say about the domain and range of such a function?
2. Find an example of such a function, where the domain of f consists of a single point.
3. Find an example of such a function on a domain consisting of two points.
4. Can such a function f exist on the integers? Why or why not?
5. Show that $(f \circ f)(x) = 1/x$ and $f(1/f(x)) = x$ for all $x \in X$.
6. Show that $f(1/x) = 1/f(x)$ for all $x \in X$.
7. Define a function g by

$$g(x) = \begin{cases} -x^3, & \text{if } x > 0 \\ -1/(x^{1/3}), & \text{if } x < 0. \end{cases}$$

Show that g satisfies $g^{-1} = 1/g$ on its domain $\mathbb{R} \setminus \{0\}$.
8. Can you find other examples of such functions?

Open-Ended Project

Here are some other common errors. Students also often confuse the composition $f \circ f$ with the product $f \cdot f$, where $(f \cdot f)(x) = f(x) \cdot f(x)$. What can you say about a function f that satisfies $f \circ f = f \cdot f$?

Yet another problem arises with powers. Which functions f have the property that $f(x^2) = (f(x))^2$ for all $x \in \text{dom}(f)$?

Notes and Sources

This project is based upon two interesting articles. The first article, [4], has several other interesting questions and problems for students. Some of them require knowledge of continuous functions. The second article, [15], is rather advanced, and it presents much more than we have here. It includes a look at complex valued functions.

27.7 Pascal's Triangle

Introduction

In this project you will explore an arithmetical triangle that was the object of study by Blaise Pascal in a treatise he wrote in 1654 (though it was known to mathematicians before him). He used this triangle to solve a question posed to him about gambling. You can find out more about the history of this problem from the references at the end of the project.

Prerequisites

This project is appropriate after Chapter 17 on induction has been covered. You should read over Problem 17.19 before you begin.

Guided Project

Pascal's triangle is presented below. Each line has one more entry than the previous line. All entries along the left and right edges of the triangle are one. Every other entry in a line is the sum of the two numbers on the line above that lie to the immediate left and right. The triangle is unbounded below.

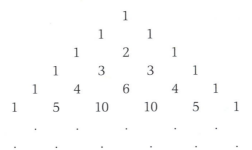

Recall that we defined n factorial and the binomial coefficient $\binom{n}{k}$ in the problems in Chapter 17.

The first few exercises should help familiarize you with Pascal's triangle.

1. Compute each of the following:

$$\binom{6}{0}, \ \binom{6}{1}, \ \binom{6}{2}, \ \binom{6}{3}, \ \binom{6}{4}, \ \binom{6}{5}, \ \text{and} \ \binom{6}{6}.$$

2. Solve Problem 17.19 (c) if you haven't already. In other words, prove that for all $k, n \in \mathbb{N}$ with $1 \le k \le n$, we get

$$\binom{n+1}{k} = \binom{n}{k-1} + \binom{n}{k}.$$

3. Use the definition of Pascal's triangle given above to show that all entries in Pascal's triangle are binomial coefficients and find a familiar mathematical expression for the kth entry from the left in the nth row. (The first entry from the left is entry 0 and the first row is row 0.) Use induction to prove that your familiar expression is correct.

4. For each $n \in \mathbb{N}$, consider the statement

$$\sum_{k=0}^{n} \binom{n}{k} = 2^{n}.$$

   (a) Check this formula for a few small values of n.
   (b) Prove the statement. (You may use Theorem 17.7.)
   (c) Show how you can obtain this sum using Pascal's triangle.

5. For each $n \in \mathbb{Z}^{+}$, consider the statement

$$\sum_{k=1}^{n} \binom{k}{k-1} = \binom{n+1}{n-1}.$$

   (a) Do something clever for a few n (as you did in 4 (a)).
   (b) Prove the statement.
   (c) Show how you can obtain this sum using Pascal's triangle.

6. How does Pascal's triangle relate to the Binomial Theorem (Theorem 17.7)?

Now you are ready for the main task of this project. Work it carefully. Be as creative, imaginative, clever, and resourceful as possible.

Open-Ended Project

Find a pattern that appears in Pascal's triangle, but that does not already appear in the text. State your formula carefully, and then prove the result. There are many different patterns!

Notes and Sources

Pascal's original article, in Latin with a French translation, appears in [63]. A very readable comprehensive history of Pascal's triangle can be found in [20].

27.8   The Cantor Set

Introduction

In this project, you'll learn about the Cantor set—a set that is in some ways very small, and in other ways very big.

Prerequisites

Proofs in this section are by induction. You will need the background provided by Chapter 17, and Chapters 20–22.

Guided Project

1. **(The Cantor Set)** To construct the Cantor set, let $I = [0, 1]$.

   (a) (First stage.) We will remove the middle third of this set; that is, we remove the open interval $(1/3, 2/3)$ from $[0, 1]$. So two intervals remain. (See Figure 27.4.) Let $E_1 = I \setminus (1/3, 2/3) = [0, 1/3] \cup [2/3, 1]$. If you were to assign a length to E_1, what length would you assign?

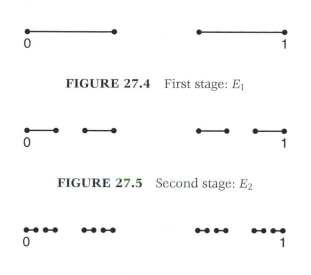

FIGURE 27.4 First stage: E_1

FIGURE 27.5 Second stage: E_2

FIGURE 27.6 Third stage: E_3

(b) (Second stage.) Remove the middle open third from each of the two remaining intervals. In other words, let $E_2 = E_1 \setminus ((1/9, 2/9) \cup (7/9, 8/9))$. So E_2 is a union of four closed intervals. (See Figure 27.5.) Write E_2 as this union of four closed intervals. If you were to assign a length to E_2, what length would you assign?

(c) (Third stage.) Remove the middle open third from each of the remaining four intervals. Thus E_3 is a union of eight intervals. (See Figure 27.6.) Write E_3 as a union of these eight closed intervals. If you were to assign a length to E_3 what length would you assign?

(d) (n^{th} stage.) Now consider E_n, obtained from E_{n-1} by removing the open middle thirds of each of the intervals that compose E_{n-1}. If you were to assign a length to E_n what length would you assign? State your guess for the length of E_n in a complete, coherent sentence. Prove that your guess is correct.

The **Cantor set** is the set E defined by $E = \bigcap_{n=1}^{\infty} E_n$.

2. If you were to assign a length to E what length would you assign? Why?

3. Give examples of numbers that you know are in the Cantor set; that is, give examples of numbers that are in E_n for every n.

4. *Another view of the Cantor set.* There are far more points in the Cantor set than you might think. To see this, it is best to revisit the Cantor set.

Each point x in the interval $[0, 1]$ has something called a ternary expansion.

The first digit in the ternary expansion for x, denoted x_1, is found as follows: We divide the interval $[0, 1]$ into thirds. If x lies in the first third, $[0, 1/3]$, we assign x_1 the value 0. If x lies in the middle third, $[1/3, 2/3]$, we assign x_1 the value 1, and if it lies in the last third, $[2/3, 1]$, we assign x_1 the value 2. (We note that there is some ambiguity about what happens at the endpoints. When working with the Cantor set (as we discuss below), whenever we have a choice, we will choose either 0 or 2 and not the number 1.)

We now proceed to the second digit, x_2, in the ternary expansion, which we find as follows: If $x_1 = 0$, then x lies in the interval $[0, 1/3]$. Divide this interval into thirds. If x lies in the first third, $[0, 1/9]$, we assign x_2 the value 0. If x lies in the middle third, $[1/9, 2/9]$, we assign x_2 the value 1. And if x lies in the final third, $[2/9, 3/9]$, we assign x_2 the value 2. Similarly, if $x_1 = 1$, then x lies in the interval $[1/3, 2/3]$, and we assign x_2 a value of 0 if x lies in the interval $[3/9, 4/9]$, a value of 1 if x lies in the interval $[4/9, 5/9]$, and a value of 2 if x lies in the interval $[5/9, 6/9]$. Finally, if $x_1 = 2$, then x lies in $[2/3, 1]$, and we divide this interval into thirds, assigning x_2 the value of 0, 1, or 2.

For x_3, we use x_1 and x_2 to tell us which interval to look at. We then divide that interval into thirds, and we assign a value of 0, 1, or 2 to x_3. It should be clear that all endpoints will have two possible representations, while all other points will have exactly one representation. Comparing the procedure defined in part 1 of this project, with the procedure we have outlined to find the ternary expansion of x, we see that the Cantor set consists of all points for which there exists a ternary expansion consisting of 0's and 2's. (That's why we never chose the number 1.) Without going into too many details, the ternary expansion really means that

$$x = \sum_{k=1}^{\infty} \frac{x_k}{3^k}, \text{ where } x_k = 0, 1, \text{ or } 2.$$

  (a) There are two ternary representations for the number 1/3.
What are they?

  (b) There are two ternary representations for the number 2/3.
What are they?

  (c) Find the first six terms of the sequence associated with 1/4.

  (d) Find the first six terms of the sequence associated with 1/8.

5. If you have studied series, then you recall that for $-1 < r < 1$, we have the following formula for the sum of the geometric series: $\sum_{k=1}^{\infty} r^k = r/(1-r)$. Using the first six terms of the ternary expansion for 1/4 that you determined above, guess all the other digits in the expansion. Then sum the series $\sum_{k=1}^{\infty} x_k/3^k$ to show that you have found the representation for 1/4. Does 1/4 lie in the Cantor set? What about 1/8?

6. We have presented the outline of a proof that there is a one-to-one correspondence between points in the Cantor set and sequences of 0's and 2's. Fill in the details, and use this to prove the theorem below.

Theorem 27.7.
The Cantor set is uncountable.

  So the Cantor set has "length" zero, but is an uncountable set. You can learn more about the Cantor set (much more) and the idea of length in the reference given below.

Open-Ended Project

What happens if instead of removing the middle third of the set, you remove the middle fifth? Think about other sets you can create in this way, and say as much as you are able to about them.

Notes and Sources

This topic is discussed in many textbooks. In particular, a summary of many of the interesting properties of this set can be found in [6, pp. 352–354]. For a short article (with a card trick) relating the Cantor set to fractals, see [9, pp. 114–121].

27.9 The Cauchy–Bunyakovsky–Schwarz Inequality

Introduction

In this project you'll prove two inequalities. The second of the two is the triangle inequality in $\mathbb{R}^n$, and the first is used to prove the second.

Prerequisites

What you need for this project depends upon how you prove it. You may need little to no background, other than an understanding of what $\mathbb{R}^n$ is and how you add and subtract in that space.

Guided Project

Consider two points, $x = (x_1, x_2, \ldots, x_n)$ and $y = (y_1, y_2, \ldots, y_n)$, in $\mathbb{R}^n$. Recall that $x + y = (x_1 + y_1, x_2 + y_2, \ldots, x_n + y_n)$ and $x \cdot y = x_1 y_1 + x_2 y_2 + \cdots + x_n y_n$. We'll introduce some notation that will make things neater. We'll write $x \cdot y = \sum_{j=1}^{n} x_j y_j$ and $\|x\| = \sqrt{x_1^2 + x_2^2 + \cdots + x_n^2}$. For $\lambda \in \mathbb{R}$, we write $\lambda x = (\lambda x_1, \lambda x_2, \ldots, \lambda x_n)$.

1. Get used to this notation: Let $x = (0, 1, 2)$ and $y = (-1, 2, 3)$ in $\mathbb{R}^3$. What is $x \cdot y$? What is $\|x\|$? $\|y\|$? $x - y$? $x + y$? Make up some examples in $\mathbb{R}^2$ and $\mathbb{R}^4$.
2. Keep getting used to this notation: What is $\{x \in \mathbb{R}^2 : \|x\| = 1\}$? What is $\{x \in \mathbb{R}^3 : \|x\| = 1\}$? If you fix $x \in \mathbb{R}^3$, what is $\{y \in \mathbb{R}^3 : \|x - y\| \leq 2\}$?
3. Let $x \in \mathbb{R}^n$ and $\lambda \in \mathbb{R}$. Prove that $\|\lambda x\| = |\lambda| \|x\|$.
4. Let $x, y \in \mathbb{R}^n$. Prove that $(x + y) \cdot (x + y) = \|x + y\|^2$.

5. Let $x, y \in \mathbb{R}^n$. Prove that $(x - y) \cdot (x - y) = \|x\|^2 - 2(x \cdot y) + \|y\|^2$.
6. Let $x, y \in \mathbb{R}^n$. Find a similar formula for $(x + y) \cdot (x + y)$.
7. Suppose x and y are two points in $\mathbb{R}^n$ such that $\|x\| = 1$ and $\|y\| = 1$. Prove that $|x \cdot y| \le 1$. (Problem 5 above together with the fact that $z \cdot z \ge 0$ for all $z \in \mathbb{R}^n$, should help to point you in the right direction.)
8. Let $x \in \mathbb{R}^n$. Prove that if $x \ne (0, 0, \dots, 0)$, then $\|x/\|x\|\| = 1$.
9. Let x and y be two points in $\mathbb{R}^n$. Prove that $|x \cdot y| \le \|x\| \|y\|$. (Problems 7 and 8 should be helpful here.) In many textbooks, this inequality is referred to as the Cauchy-Schwarz inequality; others call it the Cauchy-Bunyakovsky-Schwarz inequality.
10. Use Problems 4 and 9 to prove that for x and y in $\mathbb{R}^n$, the triangle inequality holds; that is, $\|x + y\| \le \|x\| + \|y\|$.
11. The Cauchy-Bunyakovsky-Schwarz inequality can be used to prove interesting inequalities about real numbers. Use it to prove the following: Let $a_1, a_2, \dots, a_n$ be real numbers. Then

$$\sum_{j=1}^{n} a_j^2 \ge (\sum_{j=1}^{n} a_j)^2 / n.$$

12. Bunyakovsky, Cauchy, and Schwarz all have their names attached to this theorem. Who proved what, and when did they prove it?

Open-Ended Project

For two points x and y in $\mathbb{R}^n$, the **line segment** joining x and y is defined by $\{z \in \mathbb{R}^n : z = \lambda x + (1 - \lambda)y$, where $\lambda \in \mathbb{R}$ and $0 \le \lambda \le 1\}$. For example, in $\mathbb{R}^2$ choose two points, say $(1, 2)$ and $(2, 4)$. Then the line segment joining these two points is the set

$$\{(\lambda + 2(1 - \lambda), 2\lambda + 4(1 - \lambda)), 0 \le \lambda \le 1\} = \{(2 - \lambda, 4 - 2\lambda) : 0 \le \lambda \le 1\},$$

which is indeed the line segment joining the two points $(1, 2)$ and $(2, 4)$. Try this out on other points, and in $\mathbb{R}^3$, and then move on to the next definition:

A nonempty set $S \subseteq \mathbb{R}^n$ is said to be **convex** if whenever $x, y \in S$, then the line segment joining x and y is in S. Investigate this definition, considering the following in your investigation. (You'll find the triangle inequality, as well as many of the exercises above, quite handy.)

1. Show that $\{x \in \mathbb{R}^n : \|x\| \leq 1\}$ is convex.
2. Give other examples of convex sets.
3. Is the union of two convex sets convex?
4. Is the intersection of two convex sets convex?
5. Now return to part 2 and see if you can come up with other interesting examples.
6. What are some other interesting questions (and answers) about convex sets?

Notes and Sources

For more information on the Cauchy-Bunyakovsky-Schwarz inequality, see the article by P. Schreiber [75]. There are also other (more clever, less intuitive) ways of proving this inequality.

27.10 Algebraic Numbers

Introduction

A real number is **algebraic** if it is the root of a polynomial

$$a_n x^n + \cdots + a_1 x + a_0 = 0,$$

for some positive integer n, where $a_0, a_1, \ldots, a_n \in \mathbb{Z}$ and $a_n \neq 0$. A real number is **transcendental** if it is not algebraic.

It's easy to think of examples of algebraic numbers: 0 is algebraic, because it is a (the) root of the polynomial $p(x) = x$; the number $1/2$ is algebraic, because it is a root of the polynomial $q(x) = 2x^2 - x$. It's much more difficult to think of a number that is not algebraic. Why? Well, suppose you have a guess that a certain real number a is transcendental. Then to prove your guess, you must show that for *every* polynomial p with integer coefficients, $p(a) \neq 0$. Before reading on, try to guess whether there are more transcendental numbers or more algebraic numbers.

In an 1874 paper Georg Cantor proved:

Theorem 27.8 (Cantor).
There are countably many algebraic numbers.

In this project, you will prove this theorem.

Prerequisites

This project requires material up to and including Chapter 22.

Guided Project

1. Start by getting familiar with the definition of an algebraic number by answering the next few questions. As you do so, you should also get an idea of what transcendental numbers are like, and you will begin to suspect some of your old numerical friends of being transcendentals.

    Come up with examples of algebraic real numbers that have not been presented in this project. Are some rational numbers algebraic? are all rational numbers algebraic? What about the irrational numbers? How would you prove that a particular number is algebraic or transcendental? Try to guess which of the following are transcendental numbers: $\sqrt{2}, 5/7, \pi$, and e. (If you think one of these numbers is algebraic, prove it. It's beyond our capabilities, at this time, to prove that the other numbers are transcendental.)

    Now you should be ready for the proof of Cantor's theorem. The proof is outlined below.
2. How might you attack the proof of Cantor's theorem? Does it remind you of anything we have done before? What?
3. Solve Problem 22.7, if you haven't already done so.
4. Solve Problem 22.10, if you haven't already done so.

5. Recall that for sets $A_1, A_2, \ldots, A_n$, the Cartesian product of these n sets is

$$A_1 \times A_2 \times \cdots \times A_n = \{(x_1, x_2, \ldots, x_n) : x_j \in A_j\}.$$

Prove the following generalization of Corollary 22.10.

Theorem 27.9.
Let $n \in \mathbb{N}$. If A_i is countable for all $i = 1, 2, \ldots, n$, then $A_1 \times A_2 \times \cdots \times A_n$ is countable.

6. Let $n \in \mathbb{N}$. Show that the set of polynomials of degree n with integer coefficients is countable.
7. Prove that the set of all polynomials with integer coefficients is countable.
8. Prove that the set of algebraic numbers is countable.
   You may use (but we aren't asking you to prove) the well-known fact that a polynomial of degree n has at most n distinct real roots.
9. Show that there exist transcendental numbers. (Suggestion: Don't try to actually find such a number; just try to show that they exist.)
10. Are the transcendental numbers countable or uncountable? Prove your answer.

You have now seen how the partition of the reals into the algebraic numbers and the transcendental numbers works. It is time to try something on your own.

Open-Ended Project

Define a property $\mathcal{P}$ of real numbers that seems to be of value to you. Let A denote the set of all reals that have property $\mathcal{P}$, and let B denote the set of all reals that do not have property $\mathcal{P}$. Then decide for each of the two sets, A and B, whether the set is countable or not. Prove all your statements. The more creative you are in defining $\mathcal{P}$, the harder it will be to prove the countability or uncountability of your sets. Here's a chance to really show all your mathematical prowess!

Notes and Sources

There is a difference between proving the existence of transcendental numbers, and showing that a particular number is transcendental. As we mention in the Spotlight: Hilbert's Seventh Problem, proofs that the numbers π and e are transcendental were given around the time of Cantor's proof. The original paper by G. Cantor [13] is written in German. A brief summary of Cantor's proof can be found in M. Kline's book [48, pp. 996–7].

27.11   The RSA Code

Introduction

Though coding theory has always been important, a giant leap forward occurred in the second half of the twentieth century, with the invention of public key cryptography. The main idea (due to W. Diffie and M. Hellman) is the concept of a trapdoor function—a function that has an inverse, but the inverse is very difficult to find. In fact, it should require so long for someone who did not invent the original function to find the inverse that, for all practical purposes, the inverse does not exist. In 1976, R. L. Rivest, A. Shamir, and L. M. Adleman succeeded in finding such a class of functions, and their idea is based upon one of the most elementary ideas in mathematics—multiplication of two numbers. (See the Spotlight: Public and Secret Research in Chapter 26.)

It turns out that if you take two very large numbers and multiply them together, a machine can quickly compute the answer. But, if you give the machine the answer and ask for two factors, the factorization will not appear in a useful amount of time. The public key system, built upon these ideas, is now known as RSA-key (after the three men who created the system). It was described in Chapter 26, but in this project you will learn the details.

Prerequisites

We assume that you worked Chapters 25 and 26 on modular arithmetic and Euler's theorem. In particular, we will refer to the description of the public code that was given at the end of Chapter 26. The notation was introduced in the chapter. You will also need a good calculator; one that is able to determine whether a number is prime, can factor an integer, and can do modular arithmetic. For some parts of this project, you will need to use Mathematica. (If you don't have access to Mathematica, you can skip the parts that require it.)

Guided Project

1. Reread the paragraphs of Chapter 26 following the proof of Euler's theorem.
2. Let's start with a small example to make sure we understand the basics of the code: Suppose we choose the two primes $p = 13$ and $q = 17$ (so that $n = pq = 13 \cdot 17$) and choose the encoding exponent $e = 11$. Use the public key (n, e) to calculate the "secret" values of $\phi(n)$ and d. Encode the following three plaintexts:

    (a) $m = 157$;
    (b) $m = 97216$;
    (c) $m = 91$.

    Decode them again to convince yourself that the method works.
3. Note that $\gcd(91, 13 \cdot 17) = 13 \neq 1$, so the hypothesis of Example 26.8 is not satisfied. It turns out that the method still works, even in this case. Let's try to see why it still works—what could go wrong? In this code, we always assume that n is the product of two primes: $n = pq$, where p and q are primes. Thus, the $\gcd(m, n)$ is p, q, or 1. If $\gcd(m, n) = 1$, then Example 26.8 applies. Prove that even if $\gcd(m, n) = p$ (or q), the decoding with exponent d still works:

Lemma 27.10.
Let $\gcd(m, n) = p$, *where* $n = pq$, p *and* q *are primes and* $0 < m < n$. *Further let* e *and* d *be integers such that* $ed \equiv 1 \pmod{\phi(n)}$. *Then* $m^{ed} \equiv m \pmod{n}$.

You will need Theorem 26.9 (appearing in the problem section) and Theorem 26.7 for this proof.

4. In practice, n must be chosen to be quite large—certainly larger than 10. Nevertheless, it may still be the case that the plaintext m may satisfy $m \geq n$. Recall that if $m \geq n$, then we have to break the integer m into parts. Here's how to do this: Choose positive integers $m_1, m_2, \ldots, m_k$ such that $m_i < n$ for $1 \leq i \leq k$ and $m = m_1|m_2|\ldots|m_k$, where the last expression denotes simple lining up of the integers in the decimal notation for m. (For example, if $m = 1578$ and $n = 16$, we can take $m_1 = 15$, $m_2 = 7$, and $m_3 = 8$. Then $m = 15|7|8$.) We will then denote the (chopped up) plaintext as $(m_1, m_2, \ldots, m_k)$ and the ciphertext as $(m_1^e \pmod{n}, m_2^e \pmod{n}, \ldots, m_k^e \pmod{n})$.

   Now you are ready for the problem: Suppose you are given the public key, $n = 2881$ and $e = 47$. The intercepted message contains the criminal's hair color. However, the message is encoded according to the rules we described in the previous paragraph. The ciphertext reads $(12, 285, 1057)$ (all integers $\pmod{2881}$). The translation from letters to integers is done by converting $a \to 01, b \to 02, \ldots, z \to 26$. Crack the code to find out the criminal's hair color.

5. If you cracked the message in the previous part, then it is obvious that this encryption is not safe. That's because the function we used in that part of the problem is not really a trapdoor function. However, it will become one if we choose our primes large enough. The bigger the primes, the harder it is to factor n (a task believed to be necessary to break the code). To get a feeling for the unequal amount of time it takes to find primes and multiply versus factoring, do the following on your calculator.

   (a) By trial and error using the calculator's prime check, find two primes of ten digits each. (Primality testing is also an interesting and important subject. Your calculator uses sophisticated algorithms to check whether an integer is prime.)

   (b) Multiply the two integers together. (Notice how quickly your calculator can do that!)

   (c) Now use the factor command to factor the number you obtained into its two primes. How long did it take?

6. To do safe encoding with the RSA method you need primes of 100 digits each. If you have access to Mathematica, download the notebook RSA-Notebook.nb from the site given below. Explore this package and use it to communicate with a classmate, creating public keys and sending messages to each other.
http://library.wolfram.com/infocenter/MathSource/1966/

Open-Ended Project

Either create a code of your own, or find a code from another book. Try your code out on a partner, compare it to RSA, and discuss the strengths and weaknesses of your code.

Notes and Sources

The original paper by R.L. Rivest, A. Shamir, and L.M. Adleman appears in [70]. A more detailed treatment can be found in the general number theory text book by K. H. Rosen [71, Chapter 7]. See [14] for a general link to cryptography resources at the web or [72] for a commercial site by RSA Security Inc.

To learn about primality testing, you can start with the Mathematics Magazine article [57] that gives a historical treatment of the subject up to the use of computers. A comprehensive treatment at the undergraduate level is contained in the text by Bressoud [11]. Also, a recent breakthrough is presented in the more advanced paper [1].

Spotlight: Hilbert's Seventh Problem

In 1900, David Hilbert presented a speech in Paris entitled "Mathematische Probleme" to the International Congress of Mathematicians. His aim was to look at the future of mathematics. His speech began with a description of what makes a problem significant. This introduction is followed by the statement and discussion of 23 problems. His speech appeared in 1900 in the Nachrichten of the Göttingen Scientific Society (more precisely, in Nachrichten von der Königlichen Gesellschaft der Wissenschaften zu Göttingen). It was translated into English, and published in 1902 in the Bulletin of the American Mathematical Society.

Information about the time leading up to and following the presentation can be found in C. Reid's biography *Hilbert* [68]. We present the English translation of the seventh problem below. Even today, it's exciting to hold a copy of this speech in your hands.

(You can find a definition of algebraic and transcendental numbers in Project 27.10.)

Hermite's arithmetical theorems on the exponential function and their extension by Lindemann are certain of the admiration of all generations of mathematicians. Thus the task at once presents itself to penetrate further along the path here entered, as A. Hurwitz has already done in two interesting papers,[1] "Ueber arithmetische Eigenschaften gewisser transzendenter Funktionen." I should like, therefore, to sketch a class of problems which, in my opinion, should be attacked as here next in order. That certain special transcendental functions, important in analysis, take algebraic values for certain algebraic arguments, seems to us particularly remarkable and worthy of thorough investigation. Indeed, we expect transcendental functions to assume, in general, transcendental values for even algebraic arguments; and, although it is well known that there exist integral transcendental functions which even have rational values for all algebraic arguments, we shall still consider it highly probable that the exponential function $e^{i\pi z}$, for example, which evidently has algebraic values for all rational arguments z, will on the other hand always take transcendental values for irrational algebraic values of the argument z. We can also give this statement a geometrical form, as follows:

---

[1] *Math. Annalen*, vols. 22, 32 (1883, 1888).

If, in an isosceles triangle, the ratio of the base angle to the angle at the vertex be algebraic but not rational, the ratio between base and side is always transcendental.

In spite of the simplicity of this statement and of its similarity to the problems solved by Hermite and Lindemann, I consider the proof of this theorem very difficult; as also the proof that

The expression α^β, for an algebraic base α and an irrational algebraic exponent β, e. g., the number $2^{\sqrt{2}}$ or $e^\pi = i^{-2i}$, always represents a transcendental or at least an irrational number.

It is certain that the solution of these and similar problems must lead us to entirely new methods and to a new insight into the nature of special irrational and transcendental numbers.[2]

Hilbert mentions Charles Hermite, who proved in 1873 that e is transcendental, and Ferdinand Lindemann, who proved in 1882 that π is transcendental [22, p. 466]. The answer to Hilbert's question was published in 1934 by Aleksandr O. Gelfond, and (independently) by Theodor Schneider in 1935. It follows from the Gelfond-Schneider theorem that $\sqrt{2}^{\sqrt{2}}$ is irrational (see Project 27.4), but there's an easier example. You can find this easier solution at [37].

Hilbert's original address can be found in [40]. The full text of the English translation is available on the web, [41]. See also [48, Chapter 25, sec. 1] and [48, p. 980]. For another view of Hilbert's problems read [27], and for a recent book on this topic see [28].

In honor of the 100-year anniversary of Hilbert's Paris address, the new century, and the new millenium, several mathematicians were asked to pose problems for the next century. Steve Smale proposed 18 problems for your century that you can find in [79]. The article [30] by Phillip Griffiths also contains a look at challenges for the future. The Clay Mathematics Institute of Cambridge, Massachusetts (CMI) selected seven problems for the new millenium. They also offer a reward of one million dollars per problem, and

---

[2]Quotation from [41, pp. 455–456], reprinted with permission from the publisher, the American Mathematical Society.

consequently have received a fair amount of publicity. More information about the Institute and the problems can be found on their website, [16], as well as in [18].

28

Appendix

28.1 Algebraic Properties of $\mathbb{R}$

We will assume that you are familiar with the following properties of $\mathbb{R}$.

If x and y are real numbers, then both $x + y$ and $x \cdot y$ are real numbers. Furthermore, addition and multiplication satisfy the following axioms:

A1. (The commutative property for addition) $x + y = y + x$ for all real numbers x and y;

A2. (The associative property for addition) $(x + y) + z = x + (y + z)$ for all real numbers x, y, and z;

A3. (Existence of additive identity) There is a unique real number 0 such that $0 + x = x$ for all $x \in \mathbb{R}$;

A4. (Existence of additive inverse) If $x \in \mathbb{R}$, then there is a unique element $-x$ such that $x + (-x) = 0$;

M1. (The commutative property for multiplication) $x \cdot y = y \cdot x$ for all real numbers x and y;

M2. (The associative property for multiplication) $(x \cdot y) \cdot z = x \cdot (y \cdot z)$ for all real numbers x, y, and z;

379

M3. (Existence of multiplicative identity) There is a unique real number 1, with $1 \neq 0$, such that $1 \cdot x = x$ for all real numbers x.

M4. (Existence of multiplicative inverse) For each nonzero real number x, there exists a unique real number x^{-1} such that $x \cdot x^{-1} = 1$;

D1. (The distributive property) $(x + y) \cdot z = x \cdot z + y \cdot z$ for all real numbers x, y, and z.

We note that this list of properties is not minimal; for example, the uniqueness of 0 follows from some of the other properties in the list.

28.2   Order Properties of $\mathbb{R}$

A set satisfying all of the properties above is called a **field**. Thus, $\mathbb{R}$ is an example of a field. In addition, $\mathbb{R}$ has an order defined on it. This means the following:

There is a subset $\mathbb{R}^+$ of $\mathbb{R} \setminus \{0\}$ satisfying:

O1. If $x, y \in \mathbb{R}^+$, then $x \cdot y \in \mathbb{R}^+$;

O2. If $x, y \in \mathbb{R}^+$, then $x + y \in \mathbb{R}^+$;

O3. For every real number x, exactly one of the following three things happens: either $x \in \mathbb{R}^+$, $-x \in \mathbb{R}^+$, or $x = 0$.

If x and y are two real numbers and $x - y \in \mathbb{R}^+$ we write $x > y$ (or $y < x$). The set $\mathbb{R}^+$ is called the **positive real numbers**. Thus $\mathbb{R}$ is a field with an order, and we call it an **ordered field**. The third property, O3, is called the **trichotomy principle**. It is not difficult to show that the results below follow from the statements A1–A4, M1–M4, D1, and O1–O3.

Theorem 28.1.
Let x, y, and z be real numbers. Then the following hold:
1. If $x < y$ and $y < z$, then $x < z$;
2. If $x < y$, then $x + z < y + z$;
3. If $x < y$ and $z > 0$, then $x \cdot z < y \cdot z$;
4. If $x < y$ and $z < 0$, then $x \cdot z > y \cdot z$;

5. *If $x \neq 0$, then $x^2 > 0$;*
6. *$1 > 0$;*
7. *If $x > 0$, then $x^{-1} > 0$.*

Proof.
We'll do the first and the sixth of these; you can prove the others.

For the proof of (1), note that $y - x \in \mathbb{R}^+$ and $z - y \in \mathbb{R}^+$. By O2 and the associative and commutative properties of addition, $(y - x) + (z - y) = z - x \in \mathbb{R}^+$. Therefore $z - x \in \mathbb{R}^+$ and $x < z$.

For the proof of (6), note that 1 is the multiplicative identity, so $1 \cdot x = x$ for all $x \in \mathbb{R}$. Taking $x = 1$, we get $1^2 = 1 \cdot 1 = 1$. Since $1 \neq 0$, the result now follows from (5). ■

28.3   Pólya's List[1]

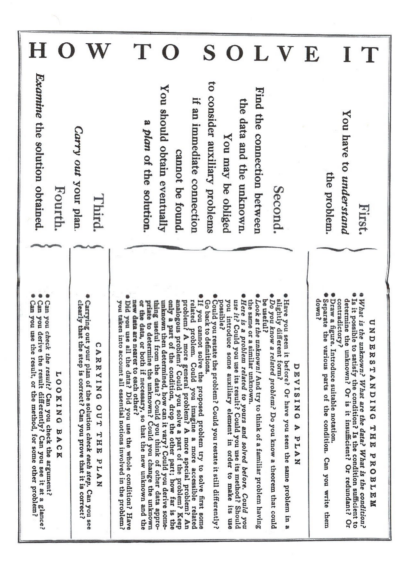

FIGURE 28.1

[1] From the inside cover of George Pólya, *How to Solve it* [66], Copyright ©1945, 1973 renewed by Princeton University Press. Reprinted by permission of Princeton University Press.

References

[1] M. Agrawal, N. Kayal, and N. Saxena. PRIMES is in P. Preprint, Department of Computer Science & Engineering, Indian Institute of Technology Kanpur, Kanpur-208016, INDIA, August 2002.

[2] D. J. Albers and G. L. Alexanderson. A conversation with Ivan Niven. *College Math. J.*, 22(5):370–402, 1991.

[3] G. L. Alexanderson. *The Random Walks of George Pólya*. Mathematical Association of America, Washington, DC, 2000.

[4] R. Anschuetz and H. Sherwood. When is a function's inverse equal to its reciprocal? *College Math. J.*, 27:388–393, 2002.

[5] D. H. Bailey, J. M. Borwein, P. B. Borwein, and S. Plouffe. The quest for pi. *Math. Intelligencer*, 19(1):50–57, 1997.

[6] R. G. Bartle and D. R. Sherbert. *Introduction to Real Analysis*. John Wiley and Sons, New York, second edition, 1992.

[7] BBC-TV/WGBH Boston co-production. *The Proof (videorecording)*, Nova Adventures in Science. South Burlington, TV: WGBH Boston Video, 1997. Produced and written by John Lynch; directed by Simon Singh.

[8] P. Beckmann. *A History of* π. The Golem Press, Boulder, CO, 1971.

[9] D. C. Benson. *The Moment of Proof*. Oxford University Press, New York, 1999.

[10] A. Bogomolny. *Cut-the-Knot.* http://www.cut-the-knot.org, (accessed April 2003).

[11] D. M. Bressoud. *Factorization and Primality Testing.* Undergraduate Texts in Mathematics. Springer-Verlag, New York, 1989.

[12] D. Burton. *Elementary Number Theory.* McGraw-Hill, Boston, MA, fifth edition, 2002.

[13] G. Cantor. Über eine Eigenschaft des Inbegriffes aller reellen algebraischen Zahlen. *J. Reine Angew. Math.*, 77:258–262, 1874.

[14] Carleton University, School of Computer Science, Ottawa. *Cryptography Resources.*
http://www.scs.carleton.ca/~csgs/resources/crypt.html, (accessed April 2003).

[15] R. Cheng, A. Dasgupta, B. R. Ebanks, L. F. Kinch, L. M. Larson, and R. B. McFadden. When does $f^{-1} = 1/f$? *Amer. Math. Monthly*, 105:704–717, 1998.

[16] Clay Mathematics Institute. *Millenium Prize Problems.*
http://www.claymath.org, (accessed April 2003).

[17] J. H. Conway and R. K. Guy. *The Book of Numbers.* Copernicus, New York, 1996.

[18] K. Devlin. *The Millennium Problems.* Basic Books, New York, 2002.

[19] W. Dunham. *Euler: The Master of Us All*, volume 22 of *The Dolciani Mathematical Expositions.* Mathematical Association of America, Washington, DC, 1999.

[20] A. W. F. Edwards. *Pascal's Arithmetical Triangle.* Oxford University Press, New York, 1987.

[21] H. M. Enzensberger. *The Number Devil: A Mathematical Adventure.* Henry Holt, New York, 1998.

[22] H. Eves. *An Introduction to the History of Mathematics.* Holt, Rinehart and Winston, New York, fourth edition, 1976.

[23] H. Eves. *Great Moments in Mathematics (after 1650)*, volume 7 of *The Dolciani Mathematical Expositions.* Mathematical Association of America, Washington, DC, 1981.

[24] C. R. Fletcher. Fermat's theorem. *Historia Math.*, 16(2):149–153, 1989.

[25] C. R. Fletcher. A reconstruction of the Frenicle–Fermat correspondence of 1640. *Historia Math.*, 18(4):344–351, 1991.

[26] L. Gårding. The Dirichlet problem. *Math. Intelligencer*, 2:43–53, 1979.

[27] I. Grattan-Guinness. A sideways look at Hilbert's twenty-three problems of 1900. *Notices Amer. Math. Soc.*, 47(7):752–757, 2000.

[28] J. J. Gray. *The Hilbert Challenge*. Oxford University Press, Oxford, 2000.

[29] Greater Online Marketing, LLC. *CalendarHome.com*. http://www.calendarhome.com/tyc/, (accessed April 2003).

[30] P. A. Griffiths. Mathematics at the turn of the millennium. *Amer. Math. Monthly*, 107(1):1–14, 2000.

[31] P. R. Halmos. *Naive Set Theory*. D. Van Nostrand Company, Inc., Princeton, NJ, 1960.

[32] P. R. Halmos. How to write mathematics. *Enseign. Math. (2)*, 16:123–152, 1970.

[33] P. R. Halmos. How to talk mathematics. *Notices Amer. Math. Soc.*, 21:155–158, 1974.

[34] P. R. Halmos. *I Want to Be a Mathematician: An Automathography*. Springer-Verlag, New York, 1985.

[35] G. H. Hardy. *A Mathematician's Apology*. Cambridge University Press, London, canto edition, 1993. First published 1940.

[36] G. H. Hardy and E. M. Wright. *An Introduction to the Theory of Numbers*. Clarendon, Oxford, first edition, 1938.

[37] Harvey Mudd College, Mathematics Department. *Math is Fun: Rational Irrational Power*. http://www.math.hmc.edu/funfacts/ffiles/10004.3-5.shtml, (accessed April 2003).

[38] T. L. Heath. *The Thirteen Books of Euclid's Elements, 3 volumes*. Dover, New York, second edition, 1956.

[39] H. Heuser. *Funktionalanalysis*. Mathematische Leitfäden. [Mathematical Textbooks]. B. G. Teubner, Stuttgart, 1992.

[40] D. Hilbert. Mathematische Probleme. *Nachrichten von der Königl. Ges. der Wiss. zu Göttingen*, pages 253–297, 1900. Also in: Archiv der Mathematik und Physik, (3) 1 (1901), pp. 44–63 and 213–237.

[41] D. Hilbert. Mathematical problems. *Bull. Amer. Math. Soc.*, 8:437–479, 1902. Translated into English by Dr. Mary Winston Newson. Full text at: http://babbage.clarku.edu/~djoyce/hilbert/problems.html (accessed April 2003).

[42] P. Høeg. *Smilla's Sense of Snow*. Farrar, Straus and Giroux, New York, 1993.

[43] A. F. Horadam. A generalized Fibonacci sequence. *Amer. Math. Monthly*, 68:455–459, 1961.

[44] D. Jarden. Curiosa: A simple proof that a power of an irrational number to an irrational exponent may be rational. *Scripta Math.*, 19:229, 1953.

[45] J. P. Jones and S. Toporowski. Irrational numbers. *Amer. Math. Monthly*, 80:423–424, 1973.

[46] V. Katz. *A History of Mathematics: An Introduction*. Addison-Wesley, Reading, MA, second edition, 1998.

[47] I. Kleiner. Evolution of the function concept: A brief survey. *College Math. J.*, 20:282–300, 1989.

[48] M. Kline. *Mathematical Thought from Ancient to Modern Times*. Oxford University Press, New York, 1972.

[49] R. Knott. *Fibonacci Numbers and the Golden Section*. http://www.mcs.surrey.ac.uk/Personal/R.Knott/Fibonacci/, (accessed April 2003).

[50] N. Koblitz. Cryptography. In B. Enquist and W. Schmid, editors, *Mathematics Unlimited—2001 and Beyond*, pages 749–769. Springer-Verlag, Berlin, 2001.

[51] S. G. Krantz. *A Primer of Mathematical Writing*. American Mathematical Society, Providence, RI, 1997.

[52] N. Luzin. Function: Part I. *Amer. Math. Monthly*, 105:59–67, 1998. Translated by Abe Shenitzer.

[53] N. Luzin. Function: Part II. *Amer. Math. Monthly*, 105:263–270, 1998. Translated by Abe Shenitzer.

[54] M. S. Mahoney. *The Mathematical Career of Pierre de Fermat, 1601–1665*. Princeton University Press, Princeton, NJ, second edition, 1994.

[55] J. E. McCarthy. How to give a good colloquium. *Canadian Mathematical Society Notes*, 31(5):3–4, 1999.

[56] E. Mendelson. *Introduction to Mathematical Logic*. Chapman & Hall, London, 1997.

[57] R. A. Mollin. A brief history of factoring and primality testing B. C. (before computers). *Math. Mag.*, 75(1):18–29, 2002.

[58] A. F. Monna. *Dirichlet's Principle—A Mathematical Comedy of Errors and Its Influence on the Development of Analysis*. Oosthoek, Scheltema and Holkema, Utrecht, the Netherlands, 1975.

[59] R. B. Nelsen. *Proofs Without Words: Exercises in Visual Thinking*. Mathematical Association of America, Washington, DC, 1993.

[60] R. B. Nelsen. *Proofs Without Words II: More Exercises in Visual Thinking*. Mathematical Association of America, Washington, DC, 2000.

[61] I. Niven. A simple proof that π is irrational. *Bull. Amer. Math. Soc. (N.S.)*, 53:509, 1947.

[62] North Dakota State University. *The Mathematics Genealogy Project*. http://genealogy.math.ndsu.nodak.edu/, (accessed, April 2003).

[63] B. Pascal. Traité du triangle arithmétique. In J. Chevalier, editor, *Oeuvres Complètes de Blaise Pascal*, Bibliothèque de la Pléiade, no. 34. Pléiade, Paris, 1954.

[64] I. Peterson. *Math Trek: The Counterfeit Coin*. http://www.maa.org/mathland/mathtrek%5F2%5F16%5F98.html, (accessed April 2003). February 1998.

[65] H. Poincaré. L'avenir des mathématiques. *Bull. des Sci. Math.*, 32:168–90, 1908.

[66] G. Pólya. *How to Solve It*. Princeton University Press, Princeton, NJ, 1945.

[67] G. Pólya. *The Pólya Picture Album: Encounters of a Mathematician*. Birkhäuser, Boston, MA, 1987. Edited by G. L. Alexanderson.

[68] C. Reid. *Hilbert*. Springer-Verlag, New York, 1970.

[69] M. von Renteln. Friedrich Prym (1841–1915)—and his investigations on the Dirichlet problem. *Rend. Circ. Mat. Palermo (2) Suppl.*, (44):43–55, 1996.

[70] R. L. Rivest, A. Shamir, and L. M. Adleman. A method for obtaining digital signatures and public-key cryptosystems. *Comm. ACM*, 21:120–126, 1978.

[71] K. H. Rosen. *Elementary Number Theory and Its Applications*. Addison-Wesley Publishing Corp., Reading, MA, third edition, 1993.

[72] RSA Security, Inc. *Company Website*. http://www.rsasecurity.com/, (accessed April 2003).

[73] D. Rüthing. Some definitions of the concept of function from Joh. Bernoulli to N. Bourbaki. *Math. Intelligencer*, 6(4):72–77, 1984.

[74] School of Mathematics and Statistics, University of St. Andrews, Scotland. *The MacTutor History of Mathematics archive*. http://www.groups.dcs.st-and.ac.uk/~history/BiogIndex.html, (accessed April 2003).

[75] P. Schreiber. The Cauchy-Bunyakovsky-Schwarz inequality. In *Hermann Graßmann (Lieschow, 1994)*, pages 64–70. Ernst-Moritz-Arndt Univ., Greifswald, 1995.

[76] P. Schumer. The Josephus problem: Once more around. *Math. Mag.*, 75:12–17, 2002.

[77] L. E. Sigler. *The Book of Squares by Leonardo Pisano Fibonacci; An Annotated Translation into Modern English*. Academic Press, Boston, MA, 1987.

[78] S. Singh. *The Code Book*. Doubleday, New York, 1999.

[79] S. Smale. Mathematical problems for the next century. *Math. Intelligencer*, 20(2):7–15, 1998.

[80] R. M. Smullyan. *What Is the Name of This Book?: The Riddle of Dracula and Other Logical Puzzles*. Prentice-Hall, Englewood Cliffs, NJ, 1978.

[81] H. Steinhaus. *Mathematical Snapshots*. Oxford University Press, New York, 1950.

[82] K. Weierstrass. Über das sogenannte Dirichlet'sche Princip, gelesen in der Königl. Akademie der Wissenschaften am 14. Juli 1870. In *Karl Weierstrass, Mathematische Werke*, volume 2, pages 49–54. Mayer & Müller, Berlin, 1895.

[83] A. Weil. *Number Theory*. Birkhäuser Boston, Inc., Boston, MA, 1984.

[84] H. Weyl. *Philosophie der Mathematik und Naturwissenschaft*. R. Oldenbourg Verlag, München, 6 edition, 1990.

[85] A. A. Wieschenberg. A conversation with George Pólya. *Math. Mag.*, 60(5):265–268, 1987.

[86] A. Wiles. Modular elliptic curves and Fermat's last theorem. *Ann. of Math. (2)*, 141(3):443–551, 1995.

[87] A. P. Youschkevitch. The concept of function up to the middle of the 19th century. *Arch. History Exact Sci.*, 16(1):37–85, 1976/77.

Index

Undergraduate Texts in Mathematics

(continued from page ii)

Franklin: Methods of Mathematical
Economics.
Frazier: An Introduction to Wavelets
Through Linear Algebra
Gamelin: Complex Analysis.
Gordon: Discrete Probability.
Hairer/Wanner: Analysis by Its History.
Readings in Mathematics.
Halmos: Finite-Dimensional Vector
Spaces. Second edition.
Halmos: Naive Set Theory.
Hämmerlin/Hoffmann: Numerical
Mathematics.
Readings in Mathematics.
Harris/Hirst/Mossinghoff:
Combinatorics and Graph Theory.
Hartshorne: Geometry: Euclid and
Beyond.
Hijab: Introduction to Calculus and
Classical Analysis.
Hilton/Holton/Pedersen: Mathematical
Reflections: In a Room with Many
Mirrors.
Hilton/Holton/Pedersen: Mathematical
Vistas: From a Room with Many
Windows.
Iooss/Joseph: Elementary Stability
and Bifurcation Theory. Second
edition.
Isaac: The Pleasures of Probability.
Readings in Mathematics.
James: Topological and Uniform
Spaces.
Jänich: Linear Algebra.
Jänich: Topology.
Jänich: Vector Analysis.
Kemeny/Snell: Finite Markov Chains.
Kinsey: Topology of Surfaces.
Klambauer: Aspects of Calculus.
Lang: A First Course in Calculus. Fifth
edition.
Lang: Calculus of Several Variables.
Third edition.
Lang: Introduction to Linear Algebra.
Second edition.
Lang: Linear Algebra. Third edition.

Lang: Short Calculus: The Original
Edition of "A First Course in
Calculus."
Lang: Undergraduate Algebra. Second
edition.
Lang: Undergraduate Analysis.
Laubenbacher/Pengelley: Mathematical
Expeditions.
Lax/Burstein/Lax: Calculus with
Applications and Computing.
Volume 1.
LeCuyer: College Mathematics with
APL.
Lidl/Pilz: Applied Abstract Algebra.
Second edition.
Logan: Applied Partial Differential
Equations.
Lovász/Pelikán/Vesztergombi: Discrete
Mathematics.
Macki-Strauss: Introduction to Optimal
Control Theory.
Malitz: Introduction to Mathematical
Logic.
Marsden/Weinstein: Calculus I, II, III.
Second edition.
Martin: Counting: The Art of
Enumerative Combinatorics.
Martin: The Foundations of Geometry
and the Non-Euclidean Plane.
Martin: Geometric Constructions.
Martin: Transformation Geometry: An
Introduction to Symmetry.
Millman/Parker: Geometry: A Metric
Approach with Models. Second
edition.
Moschovakis: Notes on Set Theory.
Owen: A First Course in the
Mathematical Foundations of
Thermodynamics.
Palka: An Introduction to Complex
Function Theory.
Pedrick: A First Course in Analysis.
Peressini/Sullivan/Uhl: The Mathematics
of Nonlinear Programming.
Prenowitz/Jantosciak: Join Geometries.

Undergraduate Texts in Mathematics